Advances in High-speed Rail Technology

"Advances in High-speed Rail Technology" presents the latest and most cutting-edge theories, techniques, and methodologies in the multidisciplinary field of high-speed railways, focusing on advances and findings from China. This series includes monographs, professional books and edited volumes from dedicated conferences and workshops. All volumes are authored or edited by established experts in their fields and undergo rigorous peer review, based on the editors' preview and selection and refereeing by independent experts. The intended audience includes researchers, engineers, industrial practitioners, graduate students, and professionals. Topics of interest in "Advances in High-speed Rail Technology" include, but are not limited to: Infrastructure, Mobile Equipment, Communication & Signal, Traction Power Supply, Operation Organization, etc.

Qizhou Hu

Natural Disaster Warning System for High-Speed Railway Safety Operation

SWJTUP 西南交通大學出版社

Springer

Qizhou Hu
Nanjing University of Science
and Technology
Nanjing, Jiangsu, China

ISSN 2363-5010 ISSN 2363-5029 (electronic)
Advances in High-speed Rail Technology
ISBN 978-981-99-7114-5 ISBN 978-981-99-7115-2 (eBook)
https://doi.org/10.1007/978-981-99-7115-2

Jointly published with Southwest Jiaotong University Press
The print edition is not for sale in China (Mainland). Customers from China (Mainland) please order the print book from: Southwest Jiaotong University Press.

This Springer imprint is published by the registered company Springer Nature Singapore Pte Ltd.
The registered company address is: 152 Beach Road, #21-01/04 Gateway East, Singapore 189721, Singapore

Preface

On October 1, 1964, the first high-speed railway (HSR) in the world was officially put into operation in Japan, which opened up a new era of transportation development. As a safe, reliable, fast, comfortable and environment-friendly transportation mode with large capacity and low-carbon emission, HSR has become a mainstream transportation mode in the development of the world's transportation industry, leading mankind to a new era. However, the safety of HSR is also a hot issue. Based on the analysis of the current situation and objective situation of HSR operation, this book explores the mechanism of HSR traffic accidents caused by natural disasters from the concept of "dynamic tracking, situation assessment and early warning management," establishes the natural disaster warning system for HSR operation safety and provides the management measures and methods for HSR operation safety under natural disasters. The natural disaster warning system for HSR operation safety contains subsystems such as crosswind warning system, rainfall warning system, temperature warning system, earthquake warning system, geological warning system and lightning warning system. They can improve the assessment and early warning capability for HSR operation safety under natural disasters, realize safe, efficient, economic, low-carbon and environment-friendly HSR operation and provide theoretical basis and technical support to improve HSR safety conditions.

This book is supported by the special fund for basic scientific research business expenses of central colleges and universities "research on key technologies for optimal allocation of traffic network resources in economic circle under high-speed rail environment" (NO. 3091601338), the key scientific and technological research project of Henan Province "intelligent decision technology for emergency traffic repair" (NO. 182102310004) and the high-level talent project of Jiangsu Province "six talent peaks" (NO. JXQC-021). We would like to express our gratitude.

This book was created by the team of Qizhou Hu, in which Xiaoyu Wu, YIkai Wu, Xin Guan, Xiang Lin, Song Ding, Longxin Zheng and other graduate students participated. Due to the limited level of the authors, there are inevitable inappropriate points in this book, and we would like to invite readers to criticize and correct them. This book can be used as a textbook for senior students and graduate students of transportation engineering, management science, system engineering, applied mathematics and other related majors in colleges and universities, or as a reference book for researchers, engineering technicians and related scholars.

Nanjing, China
May 2023

Qizhou Hu

Contents

Chapter 1
Introduction

By the end of 2022, China's HSR mileage reached 42,000 km, ranking first in the world and accounting for more than 70% of the total HSR mileage in the world. China's HSR constitutes a rapid passenger transport network with other railways, forming eight vertical and eight horizontal lines, while the high-speed railways in other countries and regions are basically in single-line operation. In particular, with the operation of Fuxing bullet trains, China's HSR has entered a new period of development. In July 2016, China released its "medium and long-term railway network planning," outlining the new period of HSR network with eight vertical and eight horizontal lines and four transnational trunk lines. By 2025, the mileage of China's HSR will reach over 50,000 km. China's HSR will connect all provincial capitals and cities with a population of over 500,000 and cover more than 90% of the country's population, achieving the goal that people can travel conveniently and goods can flow smoothly. With the rapid development of HSR, safety issues are increasingly concerned, especially HSR operation safety under the complex natural environment.

China's vast territory, complex topography, geology and diverse climate types make natural disasters more serious. There are many kinds of disasters featuring high frequency and wide geographical distribution. In particular, crosswind, heavy rain, earthquakes, debris flows, extreme temperatures, lightning and other disasters have been important factors affecting the safe operation of China's HSR. Basically, all places where HSR lines pass by are affected by natural disasters of different degrees and under the action of a trigger factor to form a mass disaster. On average, natural disasters cause more than 100 interruptions of railway transportation every year, lasting a total of 1000–2000 h, and the peak reached 211 times per year. For example, in northwest China, HSR operations face crosswind and sandstorm problems and in the northeast, there are snowstorm problems. In northeast China, HSR operations face debris flow problems, and in the southeast, there are heavy rains, etc. Figure 1.1 shows the trend of railway traffic accident rate per billion ton-kilometers from 2013 to 2022.

Q. Hu, *Natural Disaster Warning System for High-Speed Railway Safety Operation*, Advances in High-speed Rail Technology, https://doi.org/10.1007/978-981-99-7115-2_1

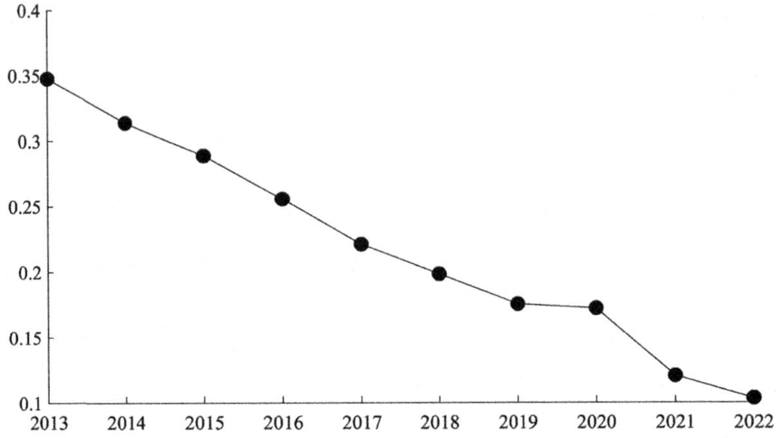

Fig. 1.1 Trend of railway traffic accident rate per billion ton-kilometers from 2013 to 2022

With the continuous improvement of HSR operation speed and grid density, HSR operations require not only the high reliability of the rolling stock, lines, power supply, and communications and signal equipment but also comprehensive and effective early warning and monitoring of various possible natural disasters (crosswind, rainstorm, earthquake, debris flow, lightning, temperature, etc.) and equipment failures to ensure safe HSR operation.

At present, various natural disasters have caused huge losses to China's railway sector and also posed a great threat to the safe and punctual operation of the HSR. For example, on July 23, 2011, motor train unit (MTU) No. D301 from Beijing South Station to Fuzhou Station and MTU No. D3115 from Hangzhou Station to Fuzhou South Station were involved in a rear-end accident due to the failure of the traction power supply catenary along the line near Wenzhou South Station caused by a lightning strike, resulting in 40 deaths and about 200 injuries. On May 13, 2014, a debris flow disaster occurred in the section from Guangzhou South Station to Shenzhen North Station due to heavy rain, resulting in the suspension of the MTUs in the section for about 9 h. On May 31, 2015, the power supply catenary was cut off due to the wind disaster, resulting in more than 10 high-speed trains in the Shenyang–Dalian section of the Harbin–Dalian high-speed railway not being able to pass normally. On June 4, 2022, the high-speed train D2809 encountered a mudslide in Guiyang–Guangzhou high-speed railway, causing one death and nine injuries. As a result, natural disasters pose a great danger to HSR operations.

1.1 Research Background

Due to the extremely high operation HSR speed, it is extremely easy to have a traffic accident when encountering a disaster. In particular, due to the frequent natural disasters in recent years, HSR safety problems have become increasingly prominent. Traffic accidents induced by disasters have caused serious social impact and economic losses. Therefore, on the basis of the existing railway system, it is an important issue faced by almost all countries to apply the traffic engineering theory and modern science and technology to improve HSR safety management and reduce traffic accidents through the study of risk definition, early warning monitoring and emergency management of HSR safety under disaster environment.

Natural disasters causing HSR safety risks are mainly extreme weather, such as heavy rain, high wind, earthquake, dust, hail, lightning and fog, etc. Therefore, the risk definition and emergency management research of HSR safety under the natural disaster environment are about grasping the action mechanism, development and change laws of accidents under the influence of various natural disasters through statistical analysis, processing and refining of the data of HSR accidents that have occurred amid natural disasters, and combining with the field simulation experiments. On this basis, a scientific assessment system is established. Then based on the assessment model and the real-time situation, we can make logical inferences in advance for accidents that are not yet known. Timely advance predictions and forecasts are made according to the degree of harm, and the corresponding warning level is determined. Finally, corresponding management measures are proposed according to the warning level.

Therefore, this book conducts an in-depth study on the current situation and macroscopic situation of HSR operation safety under natural disaster environment, studies the correlation degree between the occurrence of various traffic accidents and the types, intensity and characteristics of various disaster environments and establishes the theoretical system of risk definition and emergency management of HSR safety under natural disaster environment to improve the assessment and early warning capability of HSR operation safety, change the passive prevention of traffic safety management to active prevention and gradually form an interactive mechanism between orderly operation and traffic safety. Only in this way, can we realize the sustainable development goal of HSR, which is safe, orderly, fast, convenient and economical operation.

1.2 Foreign HSR Natural Disaster Warning Systems

The natural disaster warning system for HSR operation safety is the main system to ensure the traffic safety of HSR. The natural disaster warning system can timely monitor the natural disasters (wind, rain, snow, earthquake, geology, temperature, etc.), foreign object intrusion and emergencies that threaten HSR operation safety, collect and summarize the monitoring information of various monitoring devices and realize the distribution, centralized management and comprehensive application of monitoring information collection. In this way, we can comprehensively grasp the disaster dynamics and realize timely and accurate disaster alarm and warning. The natural disaster warning system takes corresponding emergency measures according to the severity of the disaster immediately to prevent or reduce the loss caused by the disaster, avoid secondary disasters and provide data basis for operation plans, traffic control, rescue and maintenance, etc. It is an indispensable and important technical guarantee for modern railway transportation systems.

The safety of HSR operation is an issue that railway departments around the world pay special attention to. Japan, France, Germany and other countries have always given priority to ensuring the safety of passengers' lives and property and the traffic safety and have systematically studied the safety technology as the pioneering core technology of HSR. These countries have started to plan and build the monitoring system for natural disasters in the early stage of HSR development. In view of different natural environments, geographic conditions and operating conditions, they have adopted different safety measures and continuously improved the safety countermeasures through practical application to prevent or mitigate the harm of natural disasters or emergencies to HSR operations.

1.2.1 Japan's HSR Natural Disaster Warning System

Japan is a country with frequent natural disasters such as typhoon, rainstorm, earthquake, landslide and heavy snowfall, and its railways are often subjected to natural disasters. According to statistics, about one-third of the railway accidents in Japan are caused by various natural disasters. Natural disasters seriously threaten HSR operation safety there, especially secondary disasters, which not only lead to major traffic accidents, but also cause significant economic losses. Therefore, Japan's railway department attaches great importance to the research and prevention of natural disasters. After more than 50 years of continuous research and development since the completion and operation of Shinkansen, a complete set of natural disaster prediction systems has been gradually built for simple observation, alarm and protection. The detection of natural disasters such as earthquake, high wind, rainstorm and heavy snow has been strengthened to ensure Japan's HSR operation safety. According to the types of disaster information and system functions, Japan's HSR natural disaster monitoring system can be divided into natural disaster prediction system and natural disaster detection system. Japan's natural disaster prediction system is to predict the possibility of disasters based on the monitoring data by adopting predisaster warning measures and traffic regulations to ensure safe operation. Japan's natural disaster detection system is aimed at disasters that have already happened. Through detection and judgment, the train can be prevented from entering the affected section to avoid secondary disasters. Japan's railway department has also formulated corresponding traffic safety rules in case of disasters to reduce the impact of disasters on operation and has researched and developed many automatic monitoring systems for different natural disasters, such as earthquake emergency detection and alarm systems, disaster prevention and management systems, meteorological information systems, river information systems, rail temperature monitoring systems, etc.

Currently, Shinkansen adopts a comprehensive natural disaster monitoring system (Fig. 1.2), which is based on observation devices for rain gauges, wind speed gauges, water level gauges and seismographs at corresponding locations along the line, and disaster detection devices for rock fall, landslide and debris flow, as well as rail temperature and foreign object intrusion detection devices, disaster monitoring devices for infrastructure, large buildings and stations, protective switches and protective telephones along the line. Through these devices, the system sends all kinds of disaster information along the line to the central dispatching control center and closely monitors the status of the line. In the event of a disaster, the system will automatically issue an alarm to stop the train from running and ensure the safety of Shinkansen.

Japan's disaster prevention information system adopts advanced technologies such as automatic control, monitoring, detection, alarm and satellite communication, data communication and microcomputer processing, which have greatly improved the disaster prevention capability of Shinkansen. Therefore, Shinkansen has been operating for more than 50 years with an extremely low accident rate. Japan's HSR system technically monitors the equipment status and natural disasters in real time, sets up protective projects to ensure safety, establishes a strict management system and formulates strict operation management rules under abnormal conditions. The country has also formulated and promulgated national laws to ensure HSR operation

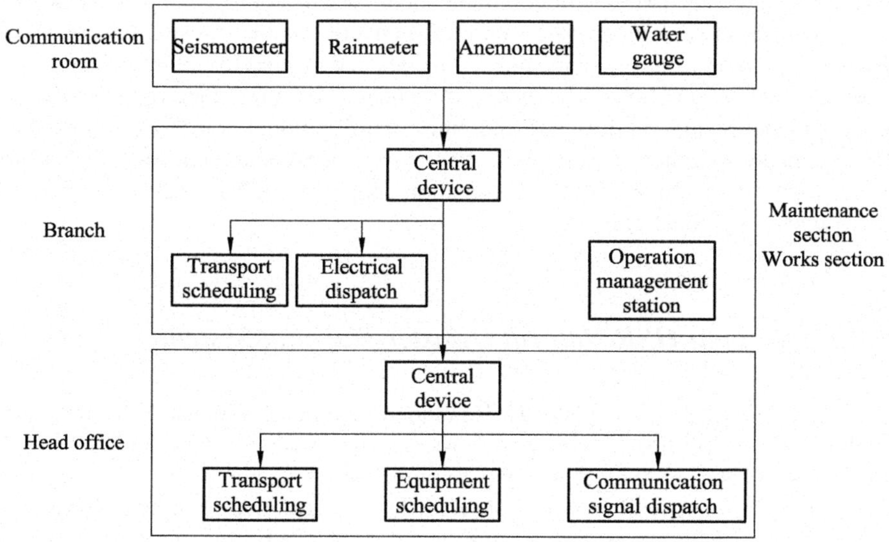

Fig. 1.2 Structure diagram of Hokkaido integrated disaster prevention information system

safety. It has been proved that Japan's railway natural disaster monitoring system is very effective. Railway accidents have been greatly reduced, and the occurrence of secondary disasters has been basically controlled.

1.2.2 French HSR Natural Disaster Warning System

French Mediterranean HSR is with a ballast track structure, operating at a speed of 300–320 km/h. Its natural disaster warning center is located in Marseille. The safety and disaster prevention monitoring equipment for crosswind, earthquake, foreign object intrusion, etc. and protective switch are set up along the line. The monitoring points are connected to the monitoring center through the communication network of SNCF. In the French Automatic Train Control (ATC) system, in addition to automatic speed control, the equipment status and natural environment detection and alarm subsystems have been added to further strengthen operation safety.

The French natural disaster monitoring system includes seven subsystem devices for automatic train detection (double failure of axle non-rotation or anti-skid system, imbalance and breakage of universal joints, and stability performance test of bogies), voltage detection of overhead lines, hot box detection, rainfall monitoring, snowfall monitoring, high wind monitoring and falling objects monitoring of overpasses. French HSR is equipped with protective switches and emergency telephones along the line. An earthquake monitoring system has been set up along the Mediterranean line by the National Seismological Bureau. French HSR and the National Seismological Bureau have jointly set up 24 unmanned seismic monitoring stations along the Mediterranean line. Two sets of communication systems, optical cable and satellite, are provided between monitoring stations to ensure reliable transmission of information. The monitoring system (Fig. 1.3) is also connected to the French National Seismic Verification Center. The seismic monitoring system is funded and used by the railway department, and designed and built by the National Seismological Bureau. The verification of the intensity level after an earthquake and the rescue after disasters are jointly conducted by the verification center of the National Seismological Bureau and the French railway.

1.2.3 German HSR Natural Disaster Warning System

German HSR is different from the HSR in Japan and France. German HSR transports both passenger and freight, and tunnels account for about one-third the total length of the line. Therefore, the traffic safety in tunnels has become the focus of its safety protection. German HSR has formulated strict and effective preventive measures. For example, freight trains without reinforcement or other protection measures and trains with dangerous goods are prohibited from entering tunnels. The intersection of passenger and freight trains in the tunnel is reduced as much as possible, and trains

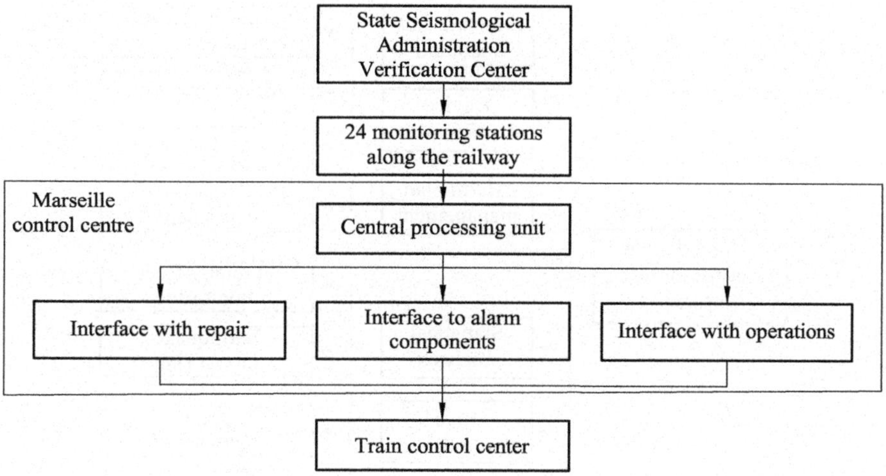

Fig. 1.3 Earthquake early warning system of Mediterranean line in France

are required to operate within a speed limit. Two trains equipped with medical and health rescue equipment have been specially manufactured for tunnel rescue, and fire brigade and rescue teams are jointly organized by local governments. In case of an accident, rescue can be carried out in time.

German HSR also adopts a new disaster prevention and alarm system (Fig. 1.4), which can not only supervise the equipment operating condition along the line but also can identify and promptly report the impact of the environment on traffic safety, as well as mobile equipment damage. The alarm system is equipped with central control units in the south, north and middle sections of the line, which are interconnected. Each central control unit is connected with various alarm and recording units located in the signal building of the main station along the line, and exchanges information and commands with them. The recording unit accepts the information collected by the detection and alarm instruments installed along the line. These detection and alarm instruments mainly include hot box detectors, tunnel airflow alarm equipment (installed in tunnels longer than 1.5 km), wind measuring instruments (installed on all bridges), fire alarm instruments, switch heating equipment, protective switches and tunnel entrance collapse alarm instruments. Emergency telephones are set at both ends of the tunnel and every 1 km inside the tunnel. The telephone box can be opened by simply wrenching the handle. Emergency calls enjoy absolute priority.

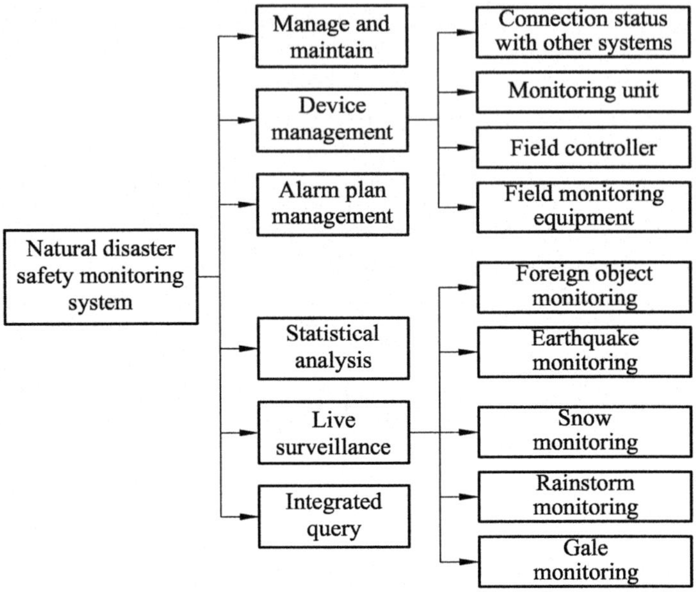

Fig. 1.4 Disaster prevention and early warning system of German high-speed railway

1.3 China's HSR Natural Disaster Warning System

China has a vast territory, and its natural conditions vary greatly in different regions with various natural disasters that occur frequently with regional and seasonal features. China's HSR operates across regions, and various natural disasters may affect HSR transportation. The main effects of natural disasters on HSR in China are as follows:

The impact of meteorological disasters on HSR: Sandstorms in northwest China and high winds in Xinjiang in spring, typhoons in the southeast coastal regions in summer and ice and snow in the northern region in winter bring inconvenience to HSR transportation.

The impact of geological disasters on HSR: The geological structure of China's southwest area is complex and prone to landslides, debris flows, etc., which affect HSR operation safety.

The impact of earthquakes on HSR: Earthquake disasters are active in some areas of China, and they are extremely sudden and destructive. It is difficult to prevent them. They bring great inconvenience to HSR operations.

China's HSR natural disaster monitoring system consists of on-site equipment for the monitoring of wind, rain, snow and foreign object intrusion, on-site monitoring units set up at Global System for Mobile Communications-Railway (GSM-R) base stations along the line, monitoring data processing equipment at each station, engineering terminals in the engineering duty room of the integrated work area of each station, dispatching equipment of each station dispatching station and transmission network, etc. Among them, the wind and rain monitoring equipment consists of anemometers, rain gauges and corresponding collection and transmission units. The foreign object intrusion monitoring equipment consists of dual power network sensors, trackside controllers and foreign object monitoring modules. The densely set monitoring points improve the reliability of the HSR natural disaster monitoring system, which is an important technical means to ensure the safe operation of China's HSR.

1.3.1 HSR Natural Disaster Warning System

The HSR natural disaster warning system consists of the disaster prevention and safety management system and the disaster prevention and safety monitoring system for passenger lines. The system exchanges and shares information with related systems about dispatching command, emergency rescue, traffic safety monitoring, passenger service, comprehensive maintenance, traction power supply, train control, China Meteorological Science Data Sharing Service Network and National Strong Earthquake Monitoring Network. A schematic diagram of the overall structure of the HSR natural disaster warning system is shown in Fig. 1.5.

China's HSR natural disaster warning system makes full use of the existing railway computer network resources. The overall network structure of the HSR natural disaster monitoring system is shown in Fig. 1.6.

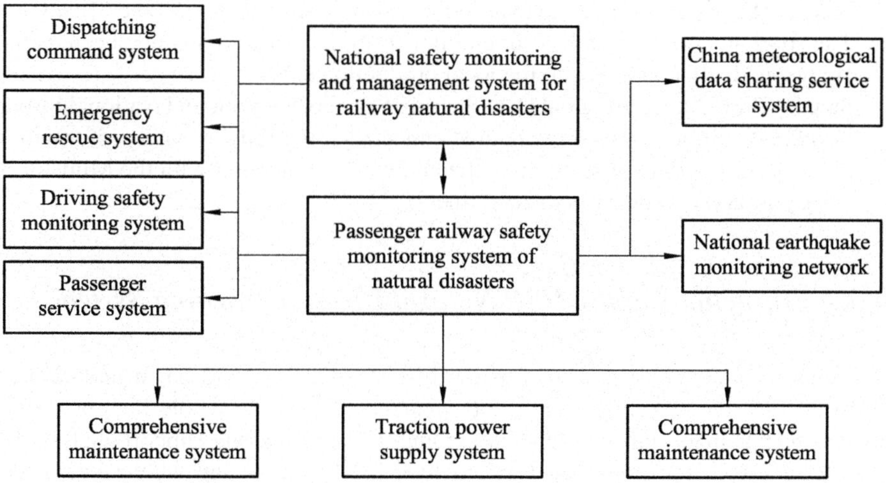

Fig. 1.5 Natural disaster warning system of high-speed railway

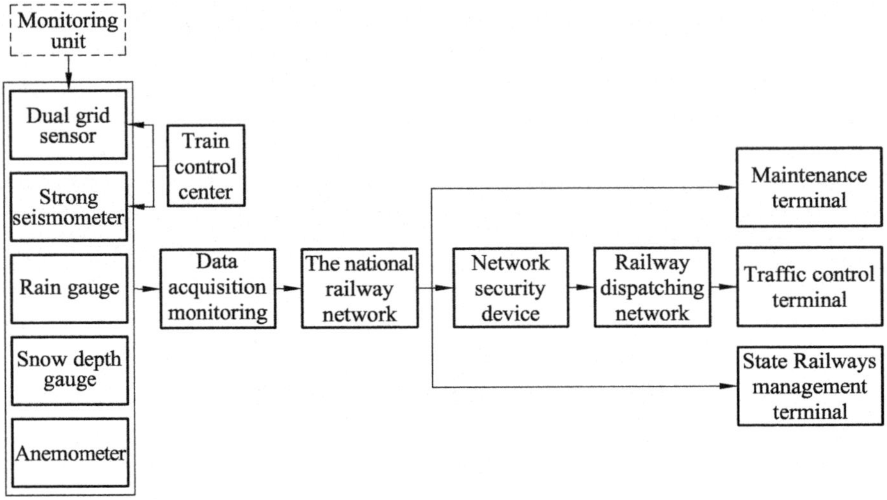

Fig. 1.6 Structure of natural disaster early warning system of high-speed railway

As can be seen from Fig. 1.6, the working process of China's HSR natural disaster warning system is as follows:

Step 1. On-site monitoring points for wind, rain, snow, earthquake and foreign object intrusion are connected to the railway computer network through dedicated lines with a network speed of 2 MB/s through the adjacent GSM-R base station and the communication machinery room of the station to realize network communication with the Ministry of Transport and passenger lines.
Step 2. The disaster prevention and safety management system of the Ministry of Transport and the natural disaster monitoring system of passenger dedicated lines are respectively connected to the local production LAN.
Step 3. China Meteorological Data Sharing Service Network and National Strong Earthquake Monitoring Network are connected to the railway safety information platform through the internet to realize network connection with the Ministry of Transport and passenger dedicated lines.

1.3.2 Functions of the HSR Natural Disaster Warning System

The disaster management system of the Ministry of Transport has a unified platform for disaster prevention and safety management on the whole line, providing macro-management, information sharing and decision-making support analysis for disaster control. The main functions include the layout of the whole monitoring network, alarm threshold setting, emergency response measures, monitoring equipment selection, application and emergency plan management, etc. The system also provides relevant basic data and monitoring data, masters disaster monitoring alarm

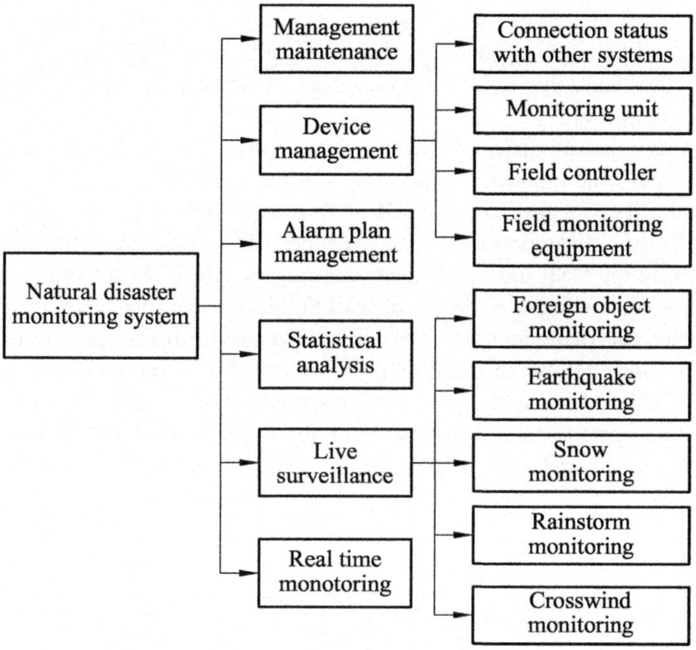

Fig. 1.7 Functions of the natural disaster early warning system for high-speed railway

and equipment operation status, supervises and guides the operation of the natural disaster monitoring system for each passenger dedicated line. Based on the analysis of disaster monitoring data of the whole line, the system provides decision-making support service for the development of the HSR natural disaster monitoring system. The functions of the HSR natural disaster warning system are shown in Fig. 1.7.

The HSR natural disaster warning system consists of four parts: on-site monitoring points along the line (wind, rain, snow, earthquake disaster and foreign object intrusion monitoring equipment), monitoring units, monitoring center and related system interfaces, etc. The system provides real-time monitoring, alarm and warning of natural disasters and emergencies, facilitates emergency response to disaster alarms, minimizes losses caused by disasters and prevents secondary disasters.

1.4 Main Warning Content

There are many kinds of natural disasters in China, but those with the greatest impact on HSR operation safety are mainly crosswind, lightning, earthquake, geology, temperature, rainstorm, etc. In order to ensure HSR operation safety under natural disasters, it is necessary to conduct risk identification and early warning management research on natural disasters. On the basis of summarizing the mechanisms of natural disasters such as crosswind, lightning, earthquake, geology, temperature and rainstorm, and by studying the conditions of domestic and foreign HSR operation monitoring systems, this book proposes a set of natural disaster warning systems suitable for China's HSR. The warning system can activate the HSR control mode in advance before natural disasters occur to slow down or stop the train, reduce losses and realize real-time monitoring and early warning. In particular, after acquiring the disaster data, the warning system will analyze it to obtain the disaster safety risk threshold. If the set safety threshold is exceeded, early warning will be started. The architecture of the HSR natural disaster warning system is shown in Fig. 1.8.

As can be seen in Fig. 1.8, the purpose of building a warning system for HSR is to take the HSR control mode in advance before disasters to ensure operation safety. The train can be slowed down or stopped to reduce disaster losses and through real-time monitoring and early warning. The early warning process is shown in Fig. 1.9.

After acquiring the natural disaster data, the disaster warning system analyzes it to obtain the safety risk threshold of the natural disaster. If the set safety threshold is exceeded, the early warning will be started. The main contents of this book are as follows:

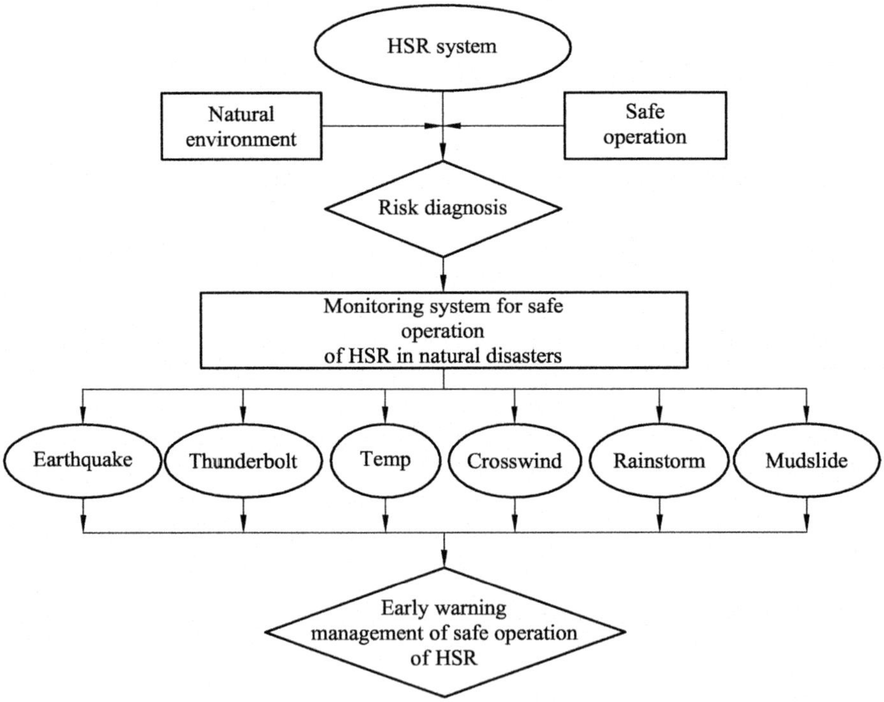

Fig. 1.8 Early warning system for safe operation of high-speed railway under natural disasters

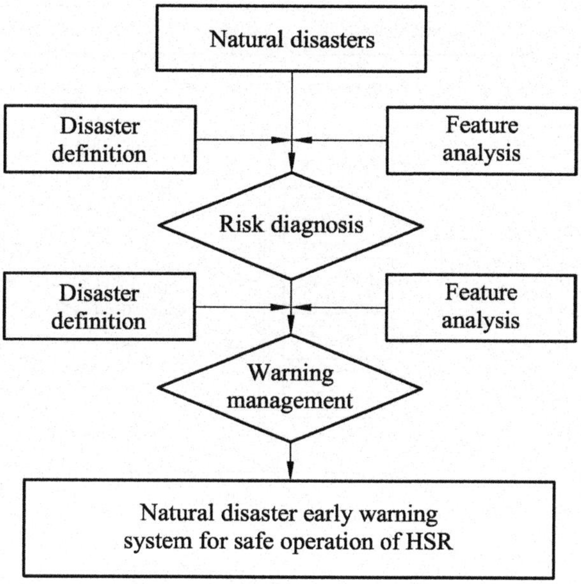

Fig. 1.9 Architecture of the early warning system

(1) Analysis of natural disaster characteristics. This book focuses on the impact of natural disasters such as crosswind, lightning, earthquake, geology, temperature and rainstorm on HSR operation safety and analyzes the characteristics of natural disasters to determine the key parameters affecting HSR operation safety.
(2) Analysis of the impact mechanism of disasters. Different natural disasters have different impact mechanisms on HSR operations. For example, the impact of crosswind on HSR is due to the asymmetry of the flow field on the windward side and the leeward side of the train. The vehicle is subjected to the action of transverse aerodynamic forces and aerodynamic moments. Earthquake causes joint action of longitudinal and transverse waves. Therefore, the impact parameters vary under different disaster conditions, and the determination of parameters is crucial.
(3) Quantitative analysis of key parameters. After determining the key parameters, according to the empirical values and different prediction models, different safety levels and warning thresholds of the parameters are defined.
(4) Construction of the natural disaster warning system for HSR operation safety. Based on analysis of natural disaster characteristics, analysis of the mechanism of disaster impact and the determination of key parameters, the early warning system for HSR operation safety under natural disasters is built.

Therefore, by analyzing the characteristics of natural disasters in China, this book studies the hazards of crosswind, lightning, earthquake, geology, extreme temperatures, rainstorm and other disasters on the safe operation of trains. Considering the actual situation of China's HSR, this book proposes the overall structure and system

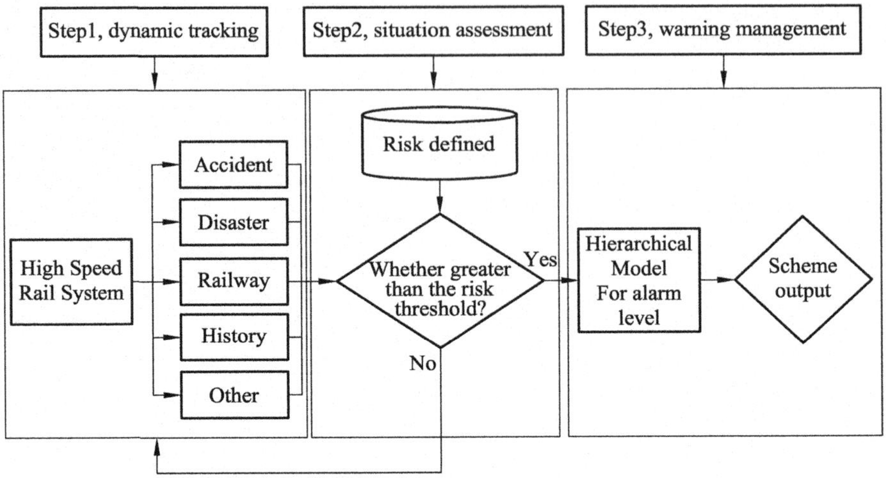

Fig. 1.10 Technical roadmap

functions of HSR safety monitoring system under natural disasters and establishes early warning methods for crosswind, lightning, earthquake, geology, temperature, rainstorm, etc. to realize real-time early warning of natural disasters that affect HSR operation safety. The research technical roadmap of this book is shown in Fig. 1.10.

Based on the actual situation of China's HSR operation safety under natural disasters (mainly crosswind, lightning, earthquake, geology, temperature, rainstorm, etc.), this book will fully absorb and learn from international advanced theories and technologies. On the basis of existing research, this book will comprehensively apply system science, mathematical method of uncertainty, nonlinear science, dynamical system theory and traffic engineering theory. By analyzing the current situation, this book will develop an early warning management method for HSR operation safety under natural disasters in light of the concept of "dynamic tracking, situation assessment and early warning management." In this way, traffic accidents can be reduced and HSR operation can be safe, efficient, economical, comfortable and less-polluting. It provides theoretical basis and technical support for improving HSR operation safety. At the same time, through statistical analysis of the existing disaster monitoring and alarm data, this book provides data support to further improve the design of the HSR system and optimize its functions.

Bibliography

1. Wang W (2014) Design and implementation of distributed real-time monitoring unit of the railway disaster prevention and safety monitoring system. Science Press, Beijing
2. Wang T(2009) Research and development of disaster prevention and safety monitoring system for high-speed railway. Chinese Railway 8:25–28

3. Wang S, Chen G (2013) Design and implementation of distributed real-time monitoring unit of the railway disaster prevention and safety monitoring system. J Lanzhou Jiaotong Univ 32(6):56–59
4. Landgraf M, Zeiner M, Knabl D, Corman F (2022) Environmental impacts and associated costs of railway turnouts based on Austrian data. Transp Res Part D Transp Environ 103:103168
5. Liu L (2010) Weather monitoring system design of high speed railway for anti-disaster. Comput Measur Control 18(9):1979–1981
6. Hirzinger B et al (2019) On the optimal strategy of stochastic-based reliability assessment of railway bridges for high-speed trains. Meccanica 54(9):1385–1402
7. Bai X, Li X, Dai X (2012) Research on the architecture of high-speed railway disaster prevention and safety monitoring system. Chinese Railway (12):27–31
8. Zhang W (2010) Railway disaster prevention and safety monitoring system for high speed railway. Railway Signalling Commun 46(6):80–81
9. Connolly et al (2020) High speed railway ground dynamics: a multi-model analysis. Int J Rail Transp 8(4):29–39
10. Li Q (2013) Construction status of railway disaster prevention and safety monitoring system and improvement measures. High Speed Railway Technol 4(6):24–28
11. Wang T, Shi H, Liu C (2009) High-speed railway disaster prevention and safety early warning system. In: The 5th China intelligent transportation annual conference and the 6th international energy-saving and new energy vehicle innovation and development—forum outstanding papers, vol 1, 25–30
12. Shen Z (2009) Design scheme of disaster protection and safety inspection system in the high-speed railway. Railway Signalling Commun Eng 6(3):6–10
13. Shi R, Li W (2008) The design of disaster prevention system for high-speed railway. J Railway Eng Soc 12(6):6–9
14. Wu Y, He F, Wang F (2012) Research on railway disaster monitoring and warning system based on disaster risk assessment model. China Railway Sci 33(1):121–125
15. Sanchis IV, Franco RI (2020) Risk of increasing temperature due to climate change on high-speed rail network in Spain. Transp Res Part D Transp Environ 82:102312

Chapter 2
An Earthquake Disaster Warning System for HSR Operation Safety

Earthquake is one of the natural disasters that threaten HSR operation safety. Human beings have limited ability to prevent earthquakes effectively. At present, the earthquake forecasting technology is not very mature. Studying earthquake early warning technology is an important measure to mitigate or avoid the danger of earthquakes to HSR operation safety. For example, in October 2004, a 6.8-magnitude earthquake occurred in Niigata, Japan. When it occurred, the Shinkansen train "Ibis 325" running between Urasa and Nagaoka derailed, as shown in Fig. 2.1. Seven cars of the 10-car train derailed, but no casualties were reported. This chapter studies the principle of earthquake warning technologies and their application in HSR, conducts a comprehensive analysis of the development status of existing HSR earthquake warning systems at home and abroad and proposes an earthquake warning system for China's HSR.

The earthquake warning system for HSR operation safety mainly focuses on such key issues as the alarm threshold, the layout of monitoring points, the train control mode and the basic composition of the warning system. Based on the experience of HSR earthquake warning in foreign countries, this chapter conducts in-depth research on the construction and operation of HSR and earthquake geological conditions in China and establishes an efficient, reliable and feasible HSR earthquake warning system. Considering China's national conditions and the characteristics of railway traffic itself, the earthquake warning system for HSR operation safety is studied through the identification and analysis of earthquake information, the interface relationship between earthquake alarm and train control system, dispatching center, station and other important nodes, and the relationship between earthquake alarm and the railway disaster prevention and safety monitoring system. This chapter focuses on the collection, analysis and recognition of earthquake monitoring information, earthquake alarm threshold, integration and information sharing between the earthquake monitoring system and the railway disaster prevention and safety monitoring system.

Q. Hu, *Natural Disaster Warning System for High-Speed Railway Safety Operation*, Advances in High-speed Rail Technology, https://doi.org/10.1007/978-981-99-7115-2_2

Fig. 2.1 Shinkansen train "Crested Ibis 325" derailed due to earthquake in 2004

2.1 Status Quo Analysis of Earthquake Early Warning System

Due to the high operation speed and density, HSR puts forward higher requirements for the traffic safety system. Countries around the world have established their HSR safety monitoring systems when building high-speed railways. These systems are mainly used to prevent and monitor natural disasters caused by wind, rain, track temperature, collapse, rockfall and earthquakes. Among the natural disasters that endanger HSR operation safety, earthquake is a sudden disaster with a small probability of occurrence but the greatest harm. When the train is running at low speed, the danger of the earthquake is not very prominent. However, since the lateral force between the wheel and track is proportional to the square of the operating speed, when the speed exceeds 200 km/h, even a small earthquake may cause a major safety accident such as train derailment or even overturning. For example, on March 5, 2010, an earthquake measuring 6.7 on the Richter scale occurred in Kaohsiung, Taiwan, causing the derailment. To prevent or mitigate the danger of earthquakes to HSR, countries and regions in the world that already have HSR have established earthquake warning systems. China is also an earthquake-prone country and is also studying earthquake warning systems for HSR operation safety.

2.1.1 *Japan's Earthquake Warning System*

Japan is an earthquake-prone country and has established many earthquake detection and warning systems. The Urgent Earthquake Detection and Alarm System (UEDAS) was developed by Japan's Railway Technical Research Institute. It works on the following principle: when an earthquake occurs, the P-wave detector set at the detection point detects the P-wave. The system deduces the magnitude, location and focal depth of the earthquake within 4 s and issues an alarm to the line section that is likely to be damaged. The seismoscope will stop, and stop the power supply within about 40 km before and after it. The train will be stopped by emergency braking, to ensure its speed is reduced to 100–170 km/h before the more harmful S-waves are propagated to Shinkansen, thus reducing the probability of possible losses or accidents. In order to deal with earthquakes in Japan's inland, a detection system with seismographs installed every 20 km along railway lines has been established. To deal with earthquakes from the Pacific Ocean, a detection system with seismographs installed at intervals of about 80 km along the coastline has been established. After the earthquake, Japan takes different measures to deal with different earthquake intensities, as shown in Table 2.1.

In addition to setting up the acceleration alarm detectors and seismographs along the line (mostly in substations), Tohoku, Joetsu and Nagano Shinkansen also set up earthquake monitoring systems along the coastline to detect earthquake waves above 40 gal (GAL is commonly used in earthquake engineering to describe the earthquake acceleration) in advance. Tokaido and Sanyo Shinkansen lines, which are close to the Tokai and Kanto earthquake areas, have adopted the more advanced "Urgent Earthquake Detection and Alarm System (UREDAS)," which uses earthquake alarms along the line (set at 40 gal) and M (magnitude)-Δ (distance from the earthquake center) to judge and control the operation control area. Figure 2.2 shows a schematic diagram of Japan's earthquake information system.

After detecting an earthquake in advance, it is essential to allow as much time as possible to slow down high-speed trains and prevent them from entering the affected area before the arrival of the S-wave. The scope and process of train operation control in the event of an earthquake are shown in Fig. 2.3.

(1) **Note 1**: In Fig. 2.3, "Earthquake intensity" is the intensity of the earthquake determined by the early monitoring system. "Exception" refers to one of the following cases.

    Case 1: An earthquake occurs when continuous rainfall reaches 120 mm or more.
    Case 2: An earthquake occurs when the track temperature rises to 50 °C or higher.
    Case 3: An earthquake occurs after sunset (including fog) (except when the intensity of the earthquake is at level C).

Table 2.1 Shinkansen railway running and inspection regulations amid earthquake

Acceleration (m/s^2)	Operating procedures	Inspection method	Remarks
< 0.8	> 30 km/h	–	After the power supply is cut, it will be repowered to resume driving
0.8–1.2	≤ 30 km/h	Patrol by train	When the road maintenance and power supply maintenance personnel patrol the train, the speed is below 70 km/h
> 1.2	Stop running	Patrol on foot	After inspection, it will run at a speed of lower than 70 km/h

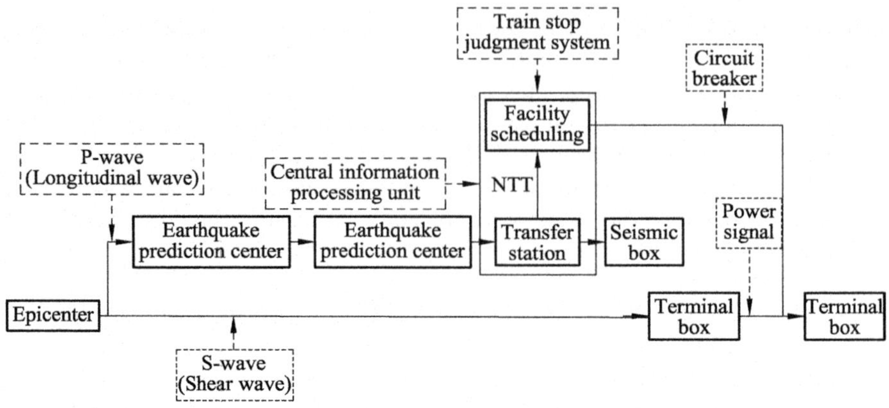

Fig. 2.2 Japan's earthquake information system

(2) **Note 2**: In Fig. 2.3, A, B, C and D are the four levels of regulations for taking corresponding measures for traffic recovery according to the relationship curve of magnitude-epicenter distance.

Rule A: Inspect the entire line after stopping.
Rule B: Inspect some sections after stopping.
Rule C: Increase speed from 70 km/h gradually after stopping.
Rule D: No stopping regulations (Table 2.2).

2.1.2 *French Earthquake Warning System*

France is not only an HSR powerhouse but also an earthquake-prone country. France has established an earthquake warning system covering all its HSR lines. In particular, the center of the earthquake monitoring network of the Mediterranean line is located in Marseille and is mainly based on the communication network of the Société nationale des chemins de fer français (SNCF). France has set up 24 monitoring points along the HSR line, with an average of 1 monitoring point every 10 km. The 24 monitoring points constitute the earthquake warning system for the Mediterranean line in France are shown in Fig. 2.4. Moreover, France has two communication networks between monitoring stations to ensure reliable transmission of earthquake information, and the system is also connected to the National Earthquake Administration Verification Center. Every earthquake monitoring point is connected to the control center in Marseille. When the center receives the alarm information from the earthquake monitoring point, it will verify it with the National Earthquake Administration Verification Center and transmit it to the train control system. Then, the operation control instruction will be issued to the train according to the set alarm threshold. The system boasts a high alarm accuracy, but there are too many signal transmission

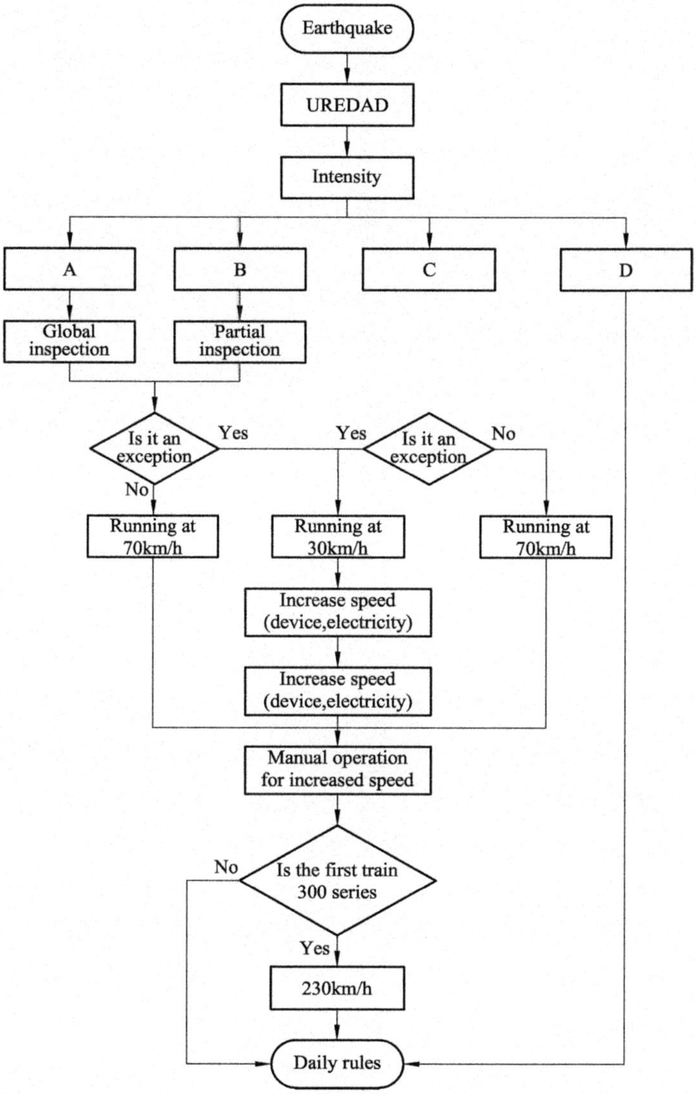

Fig. 2.3 Block diagram of the processing procedures when earthquakes occur in Japan

links and long delays. At the same time, the control center may need to manually issue train braking commands.

Unlike those in Japan, the monitoring point stations of the French Mediterranean line are not located in traction substations but are set up separately along the line. The stations are built underground with protective networks around. The advantage of being underground is to maintain the room temperature and avoid the necessity for air conditioners. There is a dual-channel network connection between the monitoring

Table 2.2 Train operation rules amid earthquake in Japan

Maximum value of seismic sensor/gal	Seismic degree	Running rules		Emergency inspection		Speed up/(km/h)		
		Stop	Speed limit	Ground inspection	Add ride inspection	30	70	> 70
[40, 80)	< 3	–	Under 70 km/h within the supervision scope, but under 30 km/h before the seismic degree is not clear, with special cases below 30 km/h	–	Nothing	–	Electrical equip.	Equip.
	> 4	–	Below 30 km/h within the scope of supervision, but there are equipment and electrical personnel the train runs under 70 km/h, especially below 30 km/h	–	Special cases are the same as the speed limit	Electrical equip.	Electrical equip.	Equip.

(continued)

Table 2.2 (continued)

Maximum value of seismic sensor/gal	Seismic degree	Running rules		Emergency inspection		Speed up/(km/h)		
		Stop	Speed limit	Ground inspection	Add ride inspection	30	70	> 70
[80, 120)	< 3	–	Below 30 km/h within the scope of supervision, but there are equipment and electrical personnel the train runs under 70 km/h, especially below 30 km/h	–	Special cases are the same as the speed limit	Electrical equip.	Electrical equip.	Equip.

(continued)

Table 2.2 (continued)

Maximum value of seismic sensor/gal	Seismic degree	Running rules		Emergency inspection		Speed up/(km/h)		
		Stop	Speed limit	Ground inspection	Add ride inspection	30	70	> 70
	4	Within the scope of supervision	Within the scope of supervision, there are equipment and electrical personnel added the train runs below 70 km/h, special cases below 30 km/h	A specific spot in the spotting area	Special cases are the same as the speed limit	Electrical equip.	Electrical equip.	Equip.

(continued)

Table 2.2 (continued)

Maximum value of seismic sensor/gal	Seismic degree	Running rules		Emergency inspection		Speed up/(km/h)		
		Stop	Speed limit	Ground inspection	Add ride inspection	30	70	> 70
	> 5	Within the scope of supervision	Within the scope of supervision, there are equipment and electrical personnel added the train runs below 70 km/h, special cases below 30 km/h	Spotting area	Special cases are the same as the speed limit	Electrical equip.	Electrical equip.	Equip.
[120, ∞)	–	Within the scope of supervision	Within the scope of supervision, there are equipment and electrical personnel added the train runs below 70 km/h, special cases below 30 km/h	Spotting area	Special cases are the same as the speed limit	Electrical equip.	Electrical equip.	Equip.

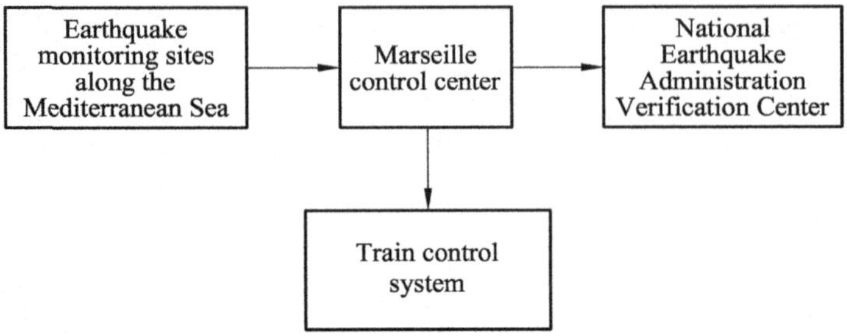

Fig. 2.4 Earthquake early warning system of the French Mediterranean line

points in France. The local monitoring system responsible for each section works independently but is connected to the central decision-making system (Marseille control center) to form an earthquake monitoring network.

The Marseille control center in France has dual redundant discriminatory processing and alarm equipment. After receiving the earthquake alarm information from three adjacent monitoring stations (if a monitoring station sends an alarm separately, the system processes it as a non-earthquake alarm message to prevent false alarms), it has to verify with the national earthquake department (the verification center is located in the IAEA). After confirmation, the information is sent to the train control system. Moreover, in Japan and France, where HSR earthquake prevention systems are relatively complete, different alarm methods are adopted in light of the specific situation, as shown in Table 2.3.

With regard to the relevant acts and emergency plans for earthquake disasters, France promulgated Decree No. 2006–1279 in August 2008, which focuses on railway transportation safety and stipulates that the overall emergency management plan must be formulated by the SNCF on a provincial basis (there are a total of 95 administrative provinces in France). The arrangement allows for a more accurate risk analysis, considering local conditions and infrastructure characteristics.

2.1.3 German Earthquake Warning System

With the rigorous working attitude of the German people, German HSR earthquake warning technology leads the world. In particular, German scientists have successfully developed a new generation of intelligent rail network system applicable to most countries and regions, which is basically the world's largest earthquake sensor. Germany's new-generation intelligent rail can continuously improve the accuracy and integrity of earthquake analysis, which is equivalent to a neural network system capable of self-debugging and learning. This intelligent rail network system can be applied in various countries and regions. The early warning system for earthquake disasters in Germany is shown in Fig. 2.5.

Table 2.3 Comparison of seismic alarm modes between Japan's Shinkansen and France's Mediterranean line

Name	Shinkansen of Japan	Mediterranean line of France
Train control mode	Send a control message to the traction power supply system to stop the power supply, and the train will brake immediately	Send a stop signal to the train control system
Alarm subject	The traction power supply system is controlled directly after the seismometer determines the authenticity of the earthquake, without the confirmation of the dispatch center	Send control information to train control system after verification by the National Verification Center according to seismometer signal along the line
Monitoring point setting	Relatively simple, set in traction substations	More complex, must build separate stations
Control signal input	Direct control, fewer transmission links, high reliability	The earthquake is verified by the National Earthquake Center, and then transmitted to the train control system through the control center. There are many transmission links
Advantages	With strong real-time performance, it can stop the train immediately when an earthquake is detected	High alarm accuracy
Disadvantages	If the discrimination is not correct, false positives may occur	The delay is long because the control center needs to manually issue the train braking command. It does not have the emergency treatment function

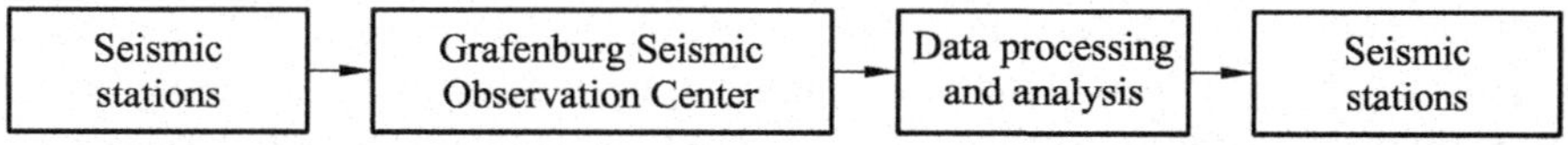

Fig. 2.5 Earthquake early warning system in Germany

To solve the problem of signal interference in the early warning system, German researchers have developed an "Embedded Rail System" (ERS), as shown in Fig. 2.6.The system has been put into practical use in some communities in Germany, Spain and the Netherlands. The vibration from street traffic near the railway, the farm tractor and the train itself can cause some interference to the sensing system, while the ERS can clearly distinguish between earthquake waves with a frequency of 30 Hz and train vibrations with a frequency of several hundred Hertz.

With regard to relevant acts and emergency plans for earthquakes, Germany has made corresponding provisions on the formulation of emergency management plans in Article 4 of its General Railway Law. Railway companies shall provide full support in the event of an emergency. In addition, they are obliged to prevent risks and investigate incidents within the railway and infrastructure areas.

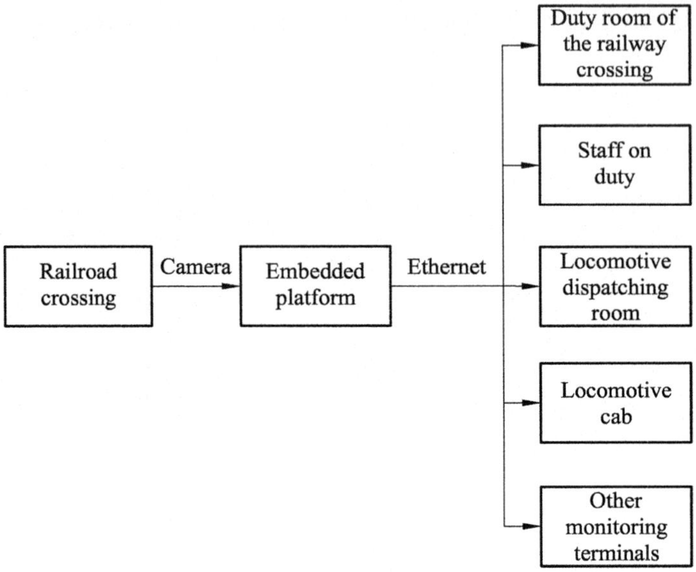

Fig. 2.6 Applications of the ERS

2.2 Earthquake Warning System for HSR

Earthquake, also known as ground motion or ground vibration, is a natural phenomenon that the earth's crust vibrates with the rapid release of energy, during which earthquake waves are generated. The main cause of earthquakes is the extrusion and collision of plates on the earth, which causes the dislocation and rupture of plate edges and internal plates. According to the concept in system engineering, the earthquake monitoring system should be studied comprehensively based on the earthquake data collection method, the layout of observation points, the identification and analysis of the earthquake information, the filtering of the interference information, the reasonable technology of the earthquake information transmission network, earthquake information sharing of multiple lines, the determination of reliability indicators, the reasonable determination of the earthquake alarm threshold, the release scope of the earthquake alarm information, the interface relationship between the earthquake alarm and the high-speed train control system, dispatching center, station and other important places, the relationship between the earthquake alarm and the current railway disaster prevention and safety monitoring system and the train deceleration or braking during an earthquake. Therefore, combined with the national conditions and the characteristics of HSR and with lessons drawn from successful foreign experience and research results, this chapter further studies the laws of earthquakes and earthquake wave propagation under different geographical and site conditions, as well as the HSR traffic safety index amid earthquakes and develops an earthquake warning system for HSR operation safety.

2.2.1 Acquisition and Identification of Earthquake Warning Information

The study on earthquake monitoring information acquisition of the HSR earthquake monitoring system includes: the technical and economic comparison among the monitoring information of China's Earthquake Networks Center, the railway self-built monitoring points (stations) and the combination of the two methods, as well as the comparison of deep monitoring and shallow monitoring methods.

(1) Early warning methods of the earthquake warning system: Earthquake parameter warning and ground motion value warning.
(2) Identification of earthquake monitoring information: Information acquisition, understanding and identification.

① Earthquake information acquisition mainly includes the verification of earthquake monitoring and management information, single-point identification and multipoint identification, the filtering and analysis of interference information.
② Earthquake information understanding: Due to the different geological structures along HSR lines, the warning information and the release scope of such information should be accurately determined.
③ Earthquake information identification: The main purpose is to improve the accuracy and speed of earthquake identification.

2.2.1.1 Earthquake Parameters

Earthquake parameters, also known as source parameters, are a quantitative representation of the earthquake source characteristics based on the analysis of the earthquake data. They include basic earthquake parameters (such as epicenter latitude and longitude, focal depth, origin time, earthquake magnitude and energy), earthquake focal mechanism solutions and focal dynamic parameters.

(1) **Classification of earthquake warning parameters according to propagation modes**. Primary wave is also called P-wave, and the transverse wave is called secondary wave (also S-wave).

① **Longitudinal wave (P-wave)**. P-wave is propulsive wave, which propagates in the earth's crust at a speed of 5.5–7 km/s and is the first to reach the epicenter. P-wave causes the ground to vibrate up and down and is less destructive.
② **Transverse wave (S-wave)**. S-wave is shear wave, which propagates in the earth's crust at a speed of 3.2–4 km/s and is the second to reach the epicenter. S-wave causes the ground to shake back and forth, left and right, and is more destructive.
③ **Surface wave (L-wave)**. L-wave is a mixed wave generated by the excitation of P-wave and S-wave meeting at the surface, with long wavelength and strong amplitude. It can only propagate along the ground surface, which is the main factor causing severe damage to buildings.

(2) **Classification of earthquake warning parameters according to earthquake waves**. Earthquake waves are mainly divided into two types. One is surface wave, which is propagated only on the surface and the other is body wave, which can travel through the interior of the earth. In addition, earthquake waves also contain surface waves, love waves and Rayleigh waves.

① **Body Wave**: It propagates inside the earth and is divided into P-wave and S-wave.

P-wave: P stands for primary or pressure and P-wave is a kind of longitudinal wave. The direction of particle vibration is parallel to the forward direction of P-wave. Among all earthquake waves, P-wave travels the fastest and arrives the first. It can also propagate in solids, liquids or gases.

S-wave: S means secondary or shear. The propagation speed of the S-wave is second only to the P-wave. The direction of particle vibration is perpendicular to the forward direction of S-wave. It is a transverse wave. Moreover, S-wave can only propagate in solids and cannot pass through liquid field core.

② **Surface Wave**: Surface wave caused by the shallow earthquake is the most obvious. Surface waves have the characteristics of low frequency, high amplitude and dispersion and propagate only near the surface. They are the most powerful seismic waves.

③ **Love Wave**: The direction of particle vibration is perpendicular to the forward direction of the wave, but the vibration only occurs in the horizontal direction without vertical component. Love waves are similar to S-waves, except that the amplitude of lateral vibration decreases with depth.

④ **Rayleigh wave**: Rayleigh waves are also known as ground roll waves. The particle motion is similar to that of sea waves. In the vertical plane, the particle vibrates in a counterclockwise ellipse, and the vibration amplitude decreases with depth.

2.2.1.2 The Mechanism of Earthquake Early Warning

Considering different propagation speeds of P-wave and S-wave, the difference in travel time between them can be used for simple earthquake localization.

(1) **Mechanisms of earthquake parameter warning**. Earthquake parameter warning uses the P-wave or S-wave detected at the station to determine the magnitude, focal depth, epicentral distance and other parameters and then determines the warning scope and level. It requires a long time to make the decision, but it is highly effective.

(2) **Mechanisms of ground motion value warning**. Ground motion value warning uses a given threshold for warning. P-wave and S-wave are not distinguished, and earthquake-related parameters are not determined, so the effectiveness is low.

(3) **Mechanisms of HSR-related earthquake warning**. HSR-related earthquake warning scheme focuses on earthquake parameter warning, and the ground

motion value warning is supplemented. It uses P-wave data for rapid determination of earthquake parameters, which is the key to earthquake warning.

2.2.1.3 Earthquake Warning Parameter Acquisition Process

The acquisition process of earthquake warning parameters is complex, involving not only the acquisition and recognition of information but also data storage. For HSR operation safety, the acquisition process is as follows:

(1) **Earthquake monitoring information acquisition for HSR operation safety**. In the acquisition of earthquake monitoring information, the front-end deep monitoring equipment uses concrete piles, which are installed in a rocky position to obtain real earthquake information and avoid the influence of ground driving, construction or other external factors on earthquake monitoring values.
(2) **Earthquake monitoring information identification for HSR operation safety**.
(3) **Earthquake information data storage for HSR operation safety**. The storage scheme of the eigenvalue of multi-level earthquake data is studied, the raw data collected by the strong-motion seismograph is stored with special codes and then eigenvalues of the data are extracted from the raw data for storage. During data query, the data is automatically selected according to the user-specified time span and retrieved from different data tables, which greatly improves the query efficiency on the basis of data accuracy.
(4) **Definition of earthquake warning parameters for HSR operation safety**. P-wave alarms have problems such as long data accumulation period, high technical difficulty and low warning accuracy. Therefore, P-wave is used to realize the HSR earthquake warning. When an earthquake is detected by the initial part of the earthquake wave, parameters such as its magnitude and location can be quickly determined.

2.2.2 *Early Warning Thresholds for Earthquakes*

Earthquake alarm thresholds should vary for different railway lines. Studying the critical values of the acceleration (horizontal acceleration and longitudinal acceleration) for operation safety under different speed target values, geological conditions, engineering facilities and different slope and curve radius. It is the key issue that needs to be studied and solved for HSR earthquake monitoring.

When the earthquake monitoring system detects the earthquake information, it is used to determine the operation management plan, which is the basis for the combination of HSR earthquake monitoring system and the operation management. The key to setting earthquake early warning thresholds is to ensure operation safety and exclude other factors (acceleration caused by the movement of the car itself). The threshold value adopted by Japan's early earthquake monitoring system is 0.045 g, which is equivalent to the level of multiple earthquakes in a seventh-degree earthquake zone. In recent years, Japan has adopted DI (structural damage index) as its monitoring target.

Where $\mathrm{DI} = \log a \cdot v + 7.0$ is the logarithm of the product of the structure acceleration a and the velocity v. The relationship between this index and earthquake intensity is the richer magnitude scale that can be described as $\mathrm{RI} = \mathrm{DI} - 0.6$.

The threshold selection path for earthquake warning systems is as follows:

Research method: Dynamic simulation analysis methods of the bridge-vehicle coupling and vehicle-line coupling are used.

Research steps: The time-histories response wave is generated by fitting the response spectrum curve given by the code of earthquake design, which acts on the bridge structure and the subgrade line after the synthesis of the track unevenness spectrum. Then the train dynamic response can be analyzed under different bridge types and slopes, subgrade heights, curve radius and train speeds which are acquired.

Research idea: Study and analyze the variation law of the derailment coefficient and the rate of wheel weight reduction. Determine the maximum allowable vertical and lateral acceleration of the train safety operation structure, as well as the ground motion acceleration value obtained with the retrodict method, which is the theoretical threshold value of the earthquake alarm. Considering domestic and foreign research and engineering application examples, the comprehensive evaluation is carried out, and the threshold standard of HSR earthquake alarm is introduced.

2.2.3 *Coupling System of Earthquake Warning System and Railway Disaster Prevention and Safety Warning System*

In the principle of reducing disaster losses and being efficient and stable, the equipment, transmission channels, power supply, grounding and response terminals and operation management strategies of the existing disaster prevention and safety monitoring system, which also includes research on emergency countermeasures in the center of the train control system, alarm methods and corresponding countermeasures in other important railway occasions such as stations, are utilized to the maximum. In this way, the coupling relationship between the earthquake monitoring system and the disaster prevention and safety monitoring system is achieved, as shown in Fig. 2.7.

At present, the monitoring objects of the HSR disaster prevention and safety monitoring system in China and abroad are mainly divided into two categories: One is for parameters that change gradually, mainly wind, rainfall and snow. The other is for mutant parameters, mainly including foreign object intrusion and earthquakes. This chapter studies HSR operation facilities, emergency control strategies, information

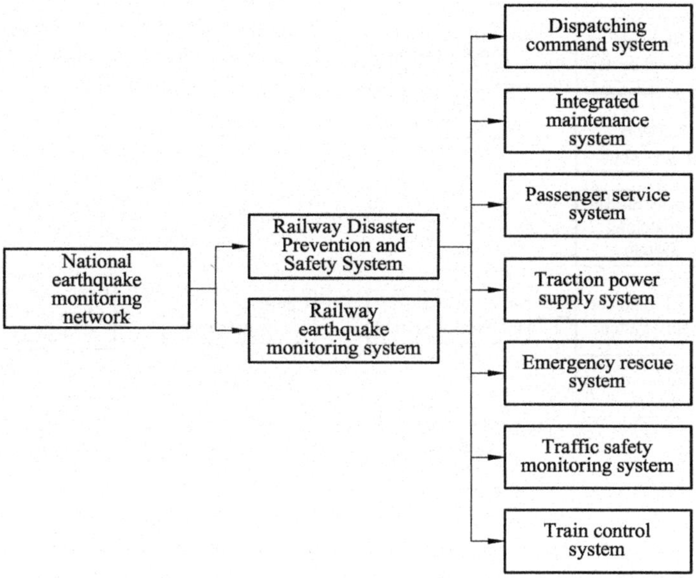

Fig. 2.7 Overall structure of HSR disaster prevention and earthquake warning system

release ranges and the linkage mode of relevant departments, and other alarm systems to build China's HSR earthquake monitoring system.

Due to the particularity of the earthquake monitoring system, its application software is relatively independent from that of other disaster prevention and safety monitoring subsystems. The call of the earthquake information database and the data storage format are also different from those of other systems. Therefore, under the principles of safety, reliability and timeliness, the HSR earthquake monitoring system should make maximum use of the server, transmission channel, power supply and grounding, disaster prevention query terminals and other units in the system to improve the comprehensive utilization rate of the earthquake monitoring system.

2.2.4 Information Sharing System for Earthquake Early Warning

On the basis of coupling between the earthquake monitoring system and the disaster prevention and safety monitoring system, the HSR earthquake monitoring system solves the problem of networking and information sharing of the multi-line earthquake monitoring system as shown in Fig. 2.8. Trains on neighboring lines or other HSR lines in the earthquake-affected area can take measures as early as possible to further improve HSR operation safety.

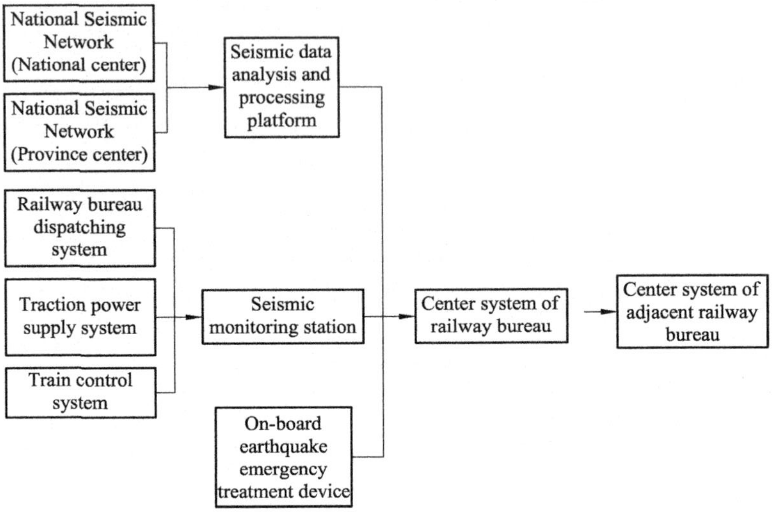

Fig. 2.8 Earthquake early warning system networking and information sharing mode

Information sharing of HSR earthquake monitoring includes external and internal sharing. External sharing is information sharing with the National Earthquake Networks, while internal sharing is information sharing among earthquake monitoring systems of multiple HSR lines. To make sure that data can be shared with the National Earthquake Networks in the near or long term, the HSR earthquake monitoring system adopts the coding format and frame structure conventions of earthquake monitoring information from National Earthquake Networks and accommodates the characteristics of the railway itself to formulate the format and coding requirements of earthquake monitoring information for the railway system.

The design and construction of the HSR earthquake monitoring system should help ensure HSR operation safety. The HSR earthquake monitoring system determines the location of earthquake stations along newly built railway lines according to the national seismic intensity table and combines the distance of the earthquake monitoring points and the direction of HSR line to comprehensively determine the location and implementation plan of the HSR earthquake monitoring system to achieve the purpose of effective monitoring. The schematic diagram of the interconnection of the HSR system and the earthquake system network through the secure data exchange area is shown in Fig. 2.9.

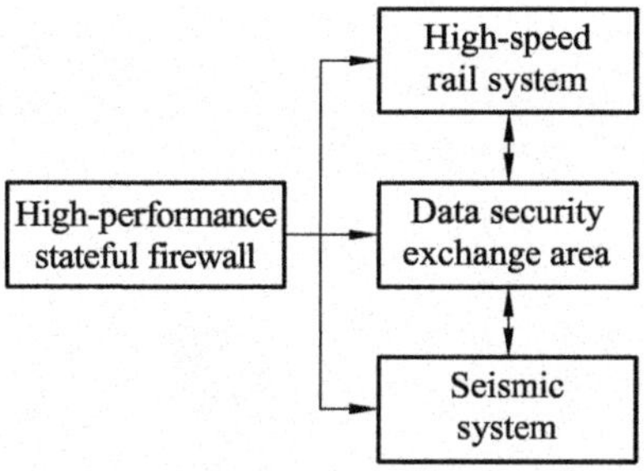

Fig. 2.9 HSR system and seismic system coupled

2.3 Influence Mechanisms of Earthquakes on HSR Operation Safety

Earthquakes generate many kinds of waves, but the main ones are P-waves and S-waves. During an earthquake, P-wave and S-wave occur at the same time. P-waves travel faster and are less destructive while S-waves travel slower, but are the main cause of damage. Therefore, the speed difference between electromagnetic waves and earthquake waves, P-waves and S-waves can be used to issue earthquake alarms within a few to dozens of seconds before an earthquake occurs and the destructive earthquake waves arrive, notifying high-speed trains to slow down or stop to avoid accidents.

A. **Classification of HSR Earthquake Warning Methods**

According to the early warning method, HSR earthquake early warning technologies are for earthquake parameter warning and the ground motion value warning respectively. Ground motion value warning is a traditional warning method, while the main development direction of the modern earthquake warning system is earthquake parameter warning.

(1) **HSR earthquake parameter warning.**
(2) **HSR ground motion value warning**. HSR ground motion value warning is directly based on whether the ground motion value exceeds a given threshold. This method is highly accurate, but it is not effective because it neither distinguishes between P-wave and S-wave nor determines relevant parameters of the earthquake in different phases.

(B) **Classification of HSR Warning Distances**

Based on the distance between the epicenter and the warning target area (the area to which an alarm needs to be issued, referring to a high-speed train), the HSR earthquake warning system can be theoretically classified into two major categories: Front-detection EWS and on-site-detection EWS.

(1) **HSR front-detection EWS**

See Fig. 2.10.

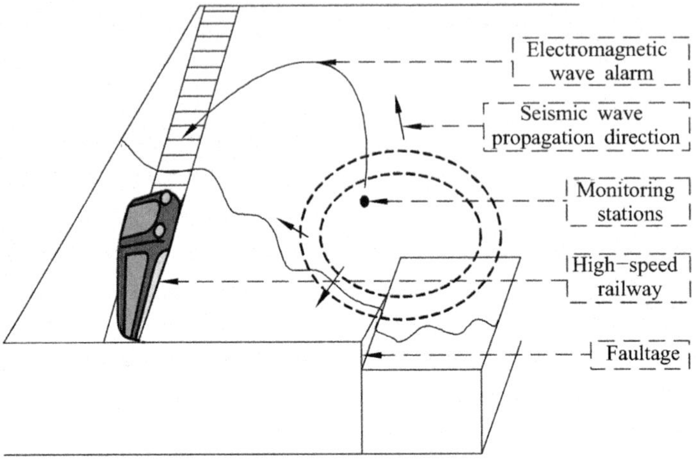

Fig. 2.10 Principles of remote earthquake early warning

(2) **HSR on-site-detection EWS**

For HSR, earthquake warning is to set up seismographs near the potential sources (front-detection EWS) or along railway lines (on-site-detection EWS). When an earthquake occurs and reaches the alarm level, the speed difference between electromagnetic waves and seismic waves or the speed difference between P-waves and S-waves is used to issue an alarm to the operating train before the arrival of destructive earthquake waves, so that the train can slow down or stop to avoid accidents. In order to improve the effectiveness of the HSR earthquake warning system, the HSR front-detection EWS and on-site-detection EWS can also be used in combination. The cost may increase, but the warning efficiency can be improved (Fig. 2.11).

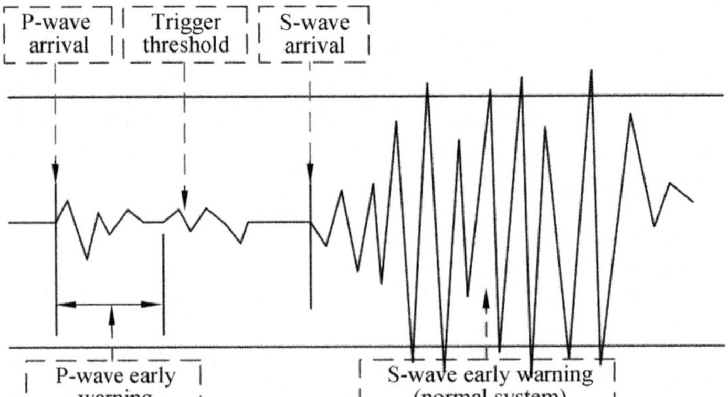

Fig. 2.11 Principles of local earthquake early warning

2.3.1 *Analysis of Earthquake Characteristics for HSR*

Earthquake waves include P-waves, S-waves and surface waves, which have different characteristics.

① **Characteristics of Earthquake P-waves.**

② **Characteristics of earthquake S-waves.** S-wave, or the transverse wave, is with a vibration direction perpendicular to the propagation direction. Its propagation speed is lower than that of P-wave. When it reaches the ground, people feel shaking and objects swing back and forth. Its propagation speed is 4–5 km/h, and it produces more destructive consequences. The propagation speed in bedrock is lower, the vibration amplitude is larger, people feel it obviously and it causes serious structural damage.

③ **Characteristics of earthquake surface waves.** When a body wave reaches the rock layer interface or the surface, it produces a wave with a large amplitude that propagates along the interface or surface, which is called a surface wave. Surface waves propagate slower than P-waves, so they follow behind S-waves. They propagate on the surface with maximum energy and the wave speed is about 3.8 km/s, which is lower than that of body waves (mainshock, $v \approx 4$ km/s).

Since P-wave propagates faster than S-wave inside the earth, P-wave always reaches the surface first during an earthquake, while S-wave always lags behind. Therefore, the seismographs in the substations along the HSR lines usually issue the alarm after the mainshock hits the line. If a high-speed train happens to run in the earthquake-affected area at the time, it is likely to fall off and overturn because it is too late to slow down. According to the experience of Japanese railways, earthquakes with a magnitude lower than M5.5 will not cause any harm to the traffic.

2.3.2 *Layout of Earthquake Monitoring Points for HSR*

The layout of HSR monitoring points should not only reduce the cost, but also meet the requirement for effective earthquake warning, so it is necessary to arrange the monitoring points rationally. There is a considerable lack of research in this area. According to the nonlinear inverse relationship between the interval of HSR earthquake monitoring points and the alarm threshold, the earthquake alarm threshold of the French Mediterranean line is 65 gal, and the average distance between earthquake monitoring points is 10 km. The earthquake alarm threshold of the Japan's Shinkansen is 40 gal, and the average distance between monitoring points is 20 km. If the alarm threshold of China's HSR is 45 gal, the interval between monitoring points should be 20–30 km, and they should be set at traction substations along the railway line.

In order to achieve a better warning effect, the monitoring points of national strong-motion observation networks can be used as peripheral monitoring points. At present,

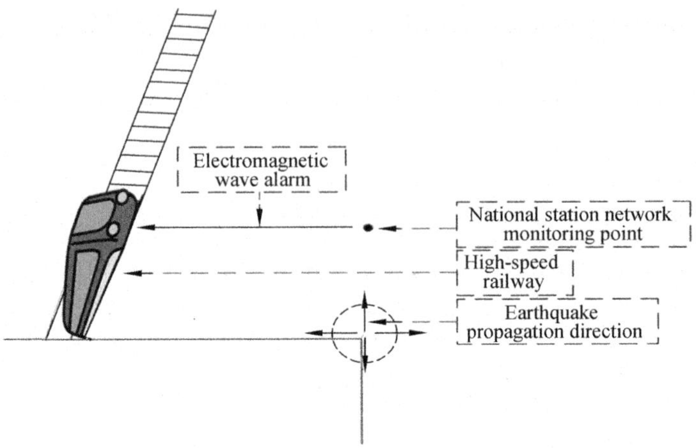

Fig. 2.12 National earthquake monitoring points as railway early warning peripheral monitoring points

the interval between the monitoring points of the strong-motion observation network established in China is more than 100 km, which cannot meet the requirements of the monitoring points for HSR earthquake warning system. However, they can be used as peripheral monitoring points of the early warning system, using the front-detection EWS to detect the earthquake earlier, as shown in Fig. 2.12.

2.3.3 Definition of HSR Earthquake Threshold

When the intensity of an earthquake reaches the set threshold, the HSR earthquake warning system issues an alarm. Indicators of the earthquake strength include earthquake intensity and ground motion acceleration (gal value).The alarm threshold of the HSR earthquake warning system is generally expressed by the gal value. Here: 1 gal = 1 cm/s^2.

A. **The function relation of the HSR earthquake alarm threshold**

The function relationship of the alarm threshold of the HSR earthquake warning system $f(\text{EAT})$ is:

$$f(\text{EAT}) = v_{[A]} \cdot [\mu_D]^{-1}, \tag{2.1}$$

where

$f(\text{EAT})$ is the alarm threshold of the HSR earthquake warning system (EAT: earthquake alarm threshold);

$v_{[A]}$ is the lateral acceleration limit of the track to ensure normal HSR operation (gal);

μ_D is the maximum dynamic response coefficient of various HSR typical structures (such as roadbeds and bridges) under the excitation of different earthquake waves, which can be determined by the dynamic response analysis of train-line-bridge or train-line coupling.

(1) **HSR earthquake warning system: the definition of μ_D.** More than 100 representative earthquake waves are selected for system analysis. The dynamic response coefficients of typical HSR bridges and standard roadbeds under the excitation of different earthquake waves are calculated. The maximum probability statistical upper value of the response coefficient under 95% guarantee rate μ_D is obtained, which is 2.55.

(2) **HSR earthquake warning system: the definition of $v_{[A]}$.** $v_{[A]}$ needs to be determined by the dynamic response analysis of the train-line-bridge coupling. The lateral acceleration limit of the HSR track. The train-line-bridge or train-line is coupled together to carry out the dynamic response analysis at the same time, and the derailment criterion is used to judge whether the high-speed train is derailed. The critical value of derailment is the lateral acceleration limit of the track.

B. Criteria for defining alarm thresholds of the HSR earthquake system

The critical value of derailment is defined mainly by derailment criteria. Through analysis and comprehensive research of domestic and foreign cases, there are mainly three criteria.

(1) **HSR derailment criterion I under the earthquake environment.** The stability against derailment of wheels is evaluated based on the lateral force Q of the HSR wheel acting on the track.

It is assumed that the wheels of the high-speed train have climbed on rail and reached the critical point (i.e., the maximum point of wheel edge inclination has been reached, as shown in Fig. 2.13), and the effect of attack angle of the wheel set and the lead of wheel-rail contact point are not considered. From the Nadal formula, it can be seen that the derailment coefficient of high-speed train wheels is

$$F(\mathrm{WDC}) = \frac{\tan\alpha - \mu}{1 + \mu\tan\alpha}, \tag{2.2}$$

where

$F(\mathrm{WDC})$ is the wheel derailment coefficient;

α is the maximum flange dip angle;

μ is the friction coefficient between the flange and the side.

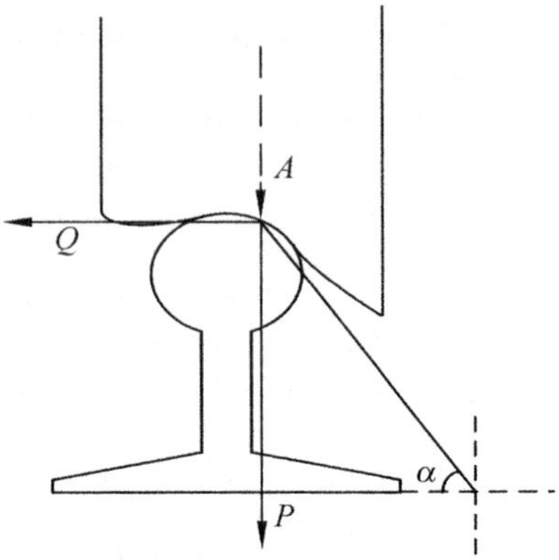

Fig. 2.13 Wheel-rail force diagram

Equation (2.2) represents the equilibrium state equation of the HSR wheel set at the critical point of the rail climbing. According to the research results in China and abroad, there is a risk of train derailment where there is an HSR wheel derailment coefficient $F(\text{WDC}) \geq 1.2$.

The warning threshold for HSR derailment: the HSR wheel derailment coefficient ≥ 1.2.

(2) **HSR derailment criterion II in an earthquake environment**

When the action time of H is longer than 0.05 s and $\mu_2 = 0.24$, the allowable value of the HSR derailment coefficient is (Fig. 2.14).

$$F(\text{AV}) = \frac{H + \mu_2 p_2}{p_1} \leq 1.2, \tag{2.3}$$

where

$F(\text{AV})$ is the allowable value;

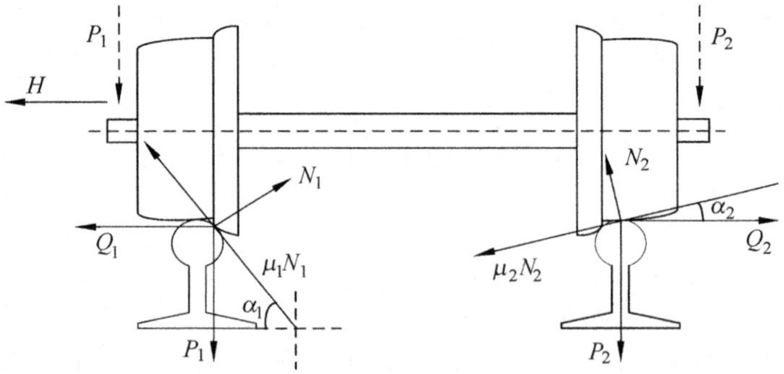

Fig. 2.14 Contact and interaction force diagram between wheel set and orbit

H is the framing force;
P_1, P_2 are the vertical forces of the wheels.

Equation (2.3) represents the equation of the allowable value of the HSR wheel set derailment coefficient. According to domestic and foreign research results, the train has derailment risks when the allowable value is greater than 1.2.

The warning threshold for HSR derailment: the allowable value of the HSR derailment coefficient $\leq$ 1.2.

(3) **HSR derailment criterion III in an earthquake environment**. The stability against derailment of the wheels is evaluated based on the rate of wheel load reduction.

The rate of wheel load reduction is

$$F(\text{WLRR}) = \frac{\Delta p}{p}, \tag{2.4}$$

where

$F(\text{WLRR})$ is the wheel load reduction rate;
P is the wheel weight;
ΔP is the wheel load reduction.

Equation (2.4) represents the state equation of the rate of wheel load reduction. According to the research results in China and abroad, when the rate of wheel load reduction is greater than 0.65, wheels of the high-speed trains are in danger of derailment.

The warning threshold of the HSR derailment: the rate of wheel load reduction of the HSR $\geq$ 0.65.

C. **Precautions for HSR earthquake alarms**

False alarms (unnecessary alarms) and missing alarms (no alarms when they are necessary) are the two major problems faced by the HSR earthquake warning system. From the perspective of passengers, the rate of false alarms should be minimized. From the perspective of safety, the rate of missing alarms should be reduced. Therefore, after the alarm threshold of the HSR earthquake warning system is determined, it should also be adjusted according to the rates of missing and false alarms during actual operation to ensure effective warning.

Due to the different geological environments along HSR lines, earthquake waves can be amplified in certain areas, while in some areas they will not be amplified, or even attenuated. Therefore, in different geological environments, the warning should be issued in specific sections to improve the accuracy of the HSR earthquake warning and minimize the inconvenience caused by false alarms.

2.3.4 Definition of the HSR Earthquake Alarm Range

The alarm modes of the HSR earthquake warning system can be roughly divided into two types. One is based on the mechanical acceleration alarm instrument (corresponding to ground motion value warning). When this seismograph monitors that the horizontal acceleration of ground motion exceeds the warning threshold, it automatically issues an alarm. The other is based on the electronic seismograph with P-wave monitoring (corresponding to ground motion parameter warning). It can estimate the distance of the earthquake epicenter from the railway line and the magnitude based on the detected P-wave and determine the necessity and scope of train control based on the magnitude and the epicenter distance. Therefore, the damage radius of the HSR (the range within which the train needs to be controlled) varies amid earthquakes of different magnitudes. How to define the HSR earthquake warning radius is one of the key issues in scientific research.

A. **Definition of the alarm radius for the HSR earthquake based on the M-R**

In order to determine the *M-R* criteria, Chinese scholar Liu Lin et al. proposed an alarm discriminant criterion with the *M-R* discriminant method. In their study, they utilized the attenuation formula of the horizontal peak acceleration of ground motion in north China proposed by scholar Peng Kezhong in 1985:

$$\lg A = -0.474 + 0.613M - 0.873 \lg R - 0.00206R, \tag{2.5}$$

where

M is the magnitude of the earthquake;
R is the epicenter distance (km);
A is the peak horizontal ground acceleration at the epicenter distance of R (km), in gal.

Under the assumption that the alarm threshold is 45 gal, the *M-R* relationship can be obtained through numerical analysis. If $A = 45$ gal and the magnitude changes from $M5.5$ to $M8.5$, the corresponding value of the epicenter distance R can be obtained. The *M-R* scatter is regressed by quadratic polynomial in the semilogarithmic coordinate system to obtain the relationship shown in Eq. (2.6). When the train is within a damage radius, an alarm is required to be issued (Fig. 2.15).

$$\lg R = -4.31774 + 1.4406M - 0.07388M^2, \tag{2.6}$$

where $M = 5.5$–8.5, which is determined according to the earthquake magnitude. The higher the magnitude, the bigger the value.

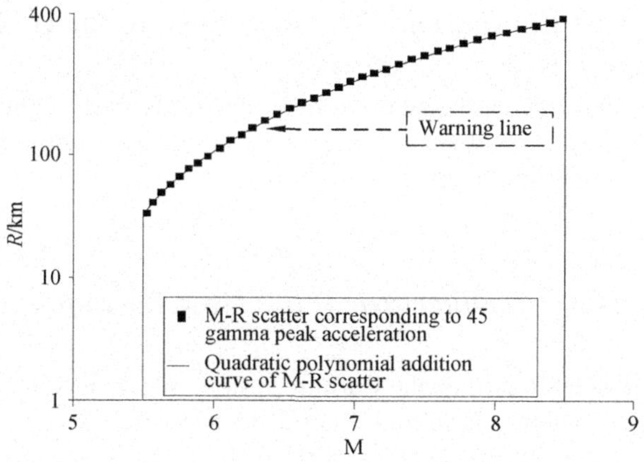

Fig. 2.15 *M-R* discriminant standard curve and warning bounds

B. **Definition of the HSR earthquake alarm threshold based on Japan's earthquake management experience**

Japan's earthquake management experience shows that earthquakes with different magnitudes and the same acceleration may cause serious damage in some cases but do not cause any damage in other cases. Therefore, the definition criteria of the HSR earthquake alarm thresholds based on Japan's earthquake management experience are as follows:

HSR earthquake alarm standard I: the HSR speed is less than 200 km/h.

HSR earthquake alarm standard II: The HSR speed is 200–350 km/h. When the speed of the high-speed train exceeds 200 km/h, the earthquake effect is more serious, so the alarm threshold decreases linearly with the increase of the train speed.

HSR earthquake alarm standard III: the HSR speed is greater than 350 km/h (Table 2.4).

Table 2.4 Train alarm thresholds

Grade interval of speed	Speed (*V*: km/h)	Alarm threshold (EAT: gal)
Interval I	< 200	60
Interval II	200–350	45
Interval III	> 350	30

2.4 Earthquake Warning System for HSR

The HSR earthquake warning system is fairly complex. This chapter constructs the HSR earthquake monitoring system by studying the structure of earthquake monitoring theory.

2.4.1 Architecture of the HSR Earthquake Warning System

The HSR earthquake warning system mainly includes data information acquisition, evaluation, alarm and control devices, as shown in Fig. 2.16. It mainly consists of components such as surface information acquisition module, storage module, communication module, monitoring host, alarm issuing module and the control module.

In Fig. 2.16, the surface information acquisition module is used to collect the basic data of the surface motion. The storage module is used to store the processed and analyzed data in the system. The evaluation and warning module is used to judge the acquired data with the warning threshold and make an evaluation. The control module is used to take different control measures according to different magnitudes after the warning judgment, and the wireless transmitting unit is used to send the warning information to the client wirelessly in time.

According to the existing research in China and abroad, the basic ground motion parameters for HSR operation control are set in Table 2.5.

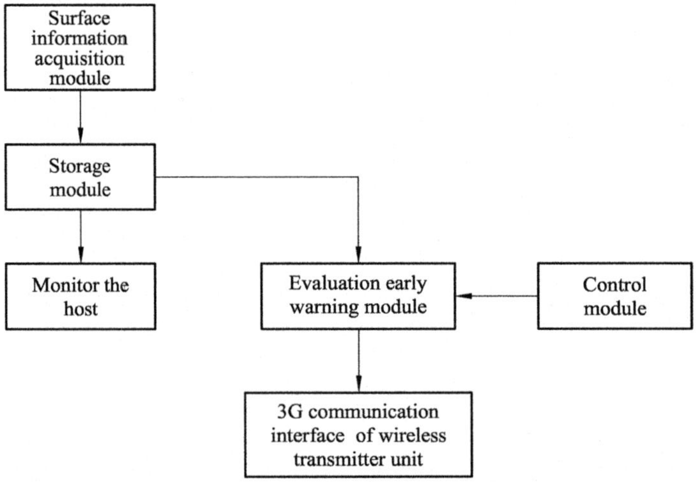

Fig. 2.16 Structure diagram of earthquake early warning devices

Table 2.5 Ground motion parameter setting for HSR operation control

Number	Name	Threshold value (gal)
1	Alarm value	40
2	Limited speed value	80
3	Stop value	120

2.4.2 HSR Earthquake Warning Process

The structure of the HSR earthquake warning system should be designed according to specific line conditions and the monitoring points should be set at stations and traction substations along the line. The earthquake monitoring point equipment consists of the seismograph and the earthquake cabinet. The seismograph includes one mechanical seismograph and an electronic seismograph to improve safety. When seismographs at the monitoring points detect the earthquake acceleration equal to or greater than the alarm threshold, they will read the information measured at the neighboring monitoring points and verify each other to avoid false alarms before issuing an earthquake alarm.

The general workflow of the HSR earthquake warning system is: earthquake monitoring points (mechanical seismograph, electronic seismograph and earthquake cabinet) → communication system (telephone line, wireless dial-up Internet and relay station) → control center (computer and special software) → alarm issuing system → train control system or traction power supply system. The basic composition of the early warning system is shown in Fig. 2.17.

A. **HSR control mode**. French control mode: The control mode of train control system represented by the French Mediterranean line is that when an alarm is received, the train control system sends a signal to control the operation and automatically stop the train.

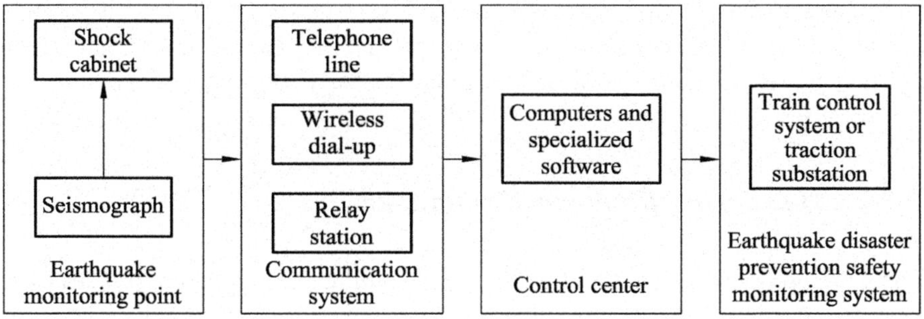

Fig. 2.17 Basic components of the earthquake early warning system

Japan's control mode: The control mode of traction substation system represented by the Japan's Shinkansen is that when an alarm is received, the traction substation stops supplying power to the train, and the high-speed train automatically slows down until it stops running due to the loss of power.

B. **Earthquake warning process**. China's earthquake warning system is not yet mature. If it is difficult to directly access the train control system with high safety requirements, it is recommended to use the control mode of traction substation system. Its HSR earthquake warning process is as follows:

Step 1: Obtain the surface motion data, calculate the earthquake parameters based on the obtained data, generate the earthquake curve and calculate the earthquake intensity at the data acquisition point.
Step 2: The system reads the preset earthquake intensity threshold and compares the calculated earthquake intensity with the threshold value. If it exceeds the threshold, the alarm process starts and the warning begins.
Step 3: According to the data stored in the device, the location of the earthquake point is obtained, and the control mode is started. If the high-speed train is controlled, the warning process ends, or the alarm process continues.

2.5 HSR Earthquake Warning Model

Earthquakes damage HSR structures, including railway lines, bridges and tunnels, which are easy to cause fatal accidents to high-speed trains. At present, China's HSR develops rapidly, but the research on earthquake monitoring systems is still in its infancy. There is a big gap with advanced countries. HSR safety technologies embody the development of modern high technology. Only by developing HSR earthquake, monitoring and warning technologies can the safety of national property and passengers be further guaranteed. China is an earthquake-prone country, and earthquakes pose a serious threat to HSR operation safety in the country. By drawing on the experience of foreign HSR earthquake warning, a comprehensive study is conducted for China's HSR operation and earthquake geological conditions to build an efficient, reliable and feasible HSR earthquake early warning model.

2.5.1 Monitoring Range of Earthquakes

The geographical distribution of earthquakes is influenced by geological structure, so there are some rules, the most obvious of which is formation of belts. The global earthquakes are mainly distributed in the following belts. One is the Circum-Pacific seismic belt, which is the most active earthquake belt in the world and 80% of the global earthquakes are concentrated there. The other is the Eurasian seismic belt, where about 15% of the global earthquakes occur. China is located in the southeastern part of the Eurasian plate and is jointly influenced by both belts, making it an earthquake-prone country. However, the spatial distribution of earthquake activities in China is highly uneven. They tend to concentrate in certain areas or zones.

The most obvious manifestations of the spatial unevenness of earthquake activities in China include not only the distribution of earthquake belts but also in homogeneity of earthquake activities in various earthquake zones. China has eight earthquake zones: the Taiwan Earthquake Zone, the Qinghai–Tibet Plateau Earthquake Zone, the Northwest China Earthquake Zone, the North China Earthquake Zone, the South China Earthquake Zone, the Northeast China Earthquake Zone, the Central China Earthquake Zone and the South China Sea Earthquake Zone. According to the actual situation of the spatial distribution of China's earthquake activities, the earthquake monitoring system for HSR operation safety is developed.

2.5.2 *Data Processing Model for HSR Earthquake Warning*

The data analysis for the HSR earthquake monitoring is mainly based on China Digital Seismic Network (CDSN), which undertakes the task of gathering and processing relevant data.

(1) **Data acquisition for HSR earthquake warning**. According to the HSR earthquake monitoring method, the key parameters for general earthquake monitoring are P-wave (speed of 7–8 km/h) and S-wave (speed of 4–5 km/h).
(2) **Data processing for HSR earthquake warning**. According to the tasks undertaken by the digital seismic network, the main contents of its data processing include seismic phase analysis, parameter determination and magnitude definition.

① **Analysis of the HSR seismic phase**. The seismic network mainly monitors local earthquakes and areas near earthquakes, so the main content of HSR seismic phase analysis is to identify the HSR seismic phase and to read the arrival time and corresponding amplitude and period.
② **Determination of HSR seismic parameters**. When an earthquake occurs, the basic parameters including the origin time of the earthquake, the latitude and longitude of epicenter and the focal depth, are determined by using specific methods through the observation data from the digital seismic network. In particular, the known conditions that must be satisfied are: location coordinates of HSR stations, velocity models or travel timetables and clear and reliable seismic phase records.
③ **Definition of HSR magnitude**. Magnitude is a measure that characterizes the strength of an earthquake. It is one of the basic parameters of earthquakes and is an important parameter in earthquake forecasting and other related seismological studies.

According to different needs, the data processing system of the digital seismic network falls into two types: one for earthquake quick report and the other for editing and publishing earthquake catalogs and observation reports. The former seeks to determine the location of earthquakes "quickly" and the positioning results are as accurate as possible, while the latter seeks to determine the location of earthquakes "precisely" and the location results must be accurate. In general, the latter is a revision of the seismic location results of the former.

(3) **Data processing mode of HSR earthquake warning**. The data processing mode of HSR earthquake warning mainly includes the calculation of earthquake origin time and earthquake azimuth.

① **Origin time of the earthquake**. Assuming that X is the signal of the earthquake wave and X_i is the amplitude of the earthquake wave at the time i, the short-time mean value of the HSR earthquake at the time n is

$$f(\text{STMV}) = \frac{1}{N_{\text{sta}}} \sum_{i=n-m}^{n} X_i, \tag{2.7}$$

where

$f(\text{STMV})$ is the short-time mean value of the earthquake and STMV is short-time mean value;

N_{sta} is the number of event points contained in short window. If the sampling time is Δt, the time length of the short window is $N_{\text{sta}} \cdot \Delta t\ k$.

Similarly, at the time n, the long-time mean value of the HSR earthquake is

$$f(\text{LTMV}) = \frac{1}{N_{\text{lta}}} \sum_{i=n-l}^{n} X_i, \tag{2.8}$$

where

$f(\text{LTMV})$ is the long-time mean value of the HSR earthquake, and LTMV is long-time mean value;

N_{lta} is the number of event points contained in long window. If the sampling time is Δt, then the time length of the long window is $N_{\text{lta}} \cdot \Delta t$.

② **Azimuth of the HSR earthquake**. The azimuth of the HSR microearthquake is the calculation of the spatial angle, which refers to the calculation of angles, which are both known and desired parameters in the three-dimensional space. Azimuth plays a key role in several aspects such as the identification of HSR microearthquake signals, correlation of the seismic phases and the event location. This chapter adopts the basic algorithm that uses vectors to solve the azimuth, which is to find the corresponding angle by the ratio of the average velocity components of each channel on the three-component sensor.

The average vibration velocity in the north–south direction of the HSR earthquake is

$$\overline{v}_n = \frac{1}{k}\sum_{i=0}^{k} F_n(i). \tag{2.9}$$

The average vibration velocity in the east–west direction of the HSR earthquake is

$$\overline{v}_e = \frac{1}{k}\sum_{i=0}^{k} F_e(i). \tag{2.10}$$

The average vibration velocity in the elevation direction of the HSR earthquake is

$$\overline{v}_z = \frac{1}{k}\sum_{i=0}^{k} F_z(i), \tag{2.11}$$

where

k	is the time span of the three-component vibration signal;
$F_n(i)$, $F_e(i)$ and $F_z(i)$	respectively represent the function of the vibration velocity as a function of time in the north–south, east–west and elevation directions;
$\overline{v}_n$, $\overline{v}_e$ and $\overline{v}_z$	denote the average vibration velocity of event points in corresponding directions, respectively.

In an HSR earthquake, the average velocity in each direction is used to obtain the corresponding azimuth through arctangent operation, and the projection angle on the horizontal plane in the HSR earthquake is

$$f(\text{azim}) = \arctan\frac{\overline{v}_e}{\overline{v}_n}, \tag{2.12}$$

where

$f(\text{azim})$ is the projection angle on the horizontal plane.

The projection angle between the vertical plane and the horizontal plane in the HSR earthquake is

$$f(\text{dip}) = \arctan\frac{\sqrt{\overline{v}_e^2 + \overline{v}_n^2}}{\overline{v}_z}, \tag{2.13}$$

where

f(dip) is the projection angle between vertical and horizontal planes.

(4) **Event location of HSR earthquake warning**. Event location in microearthquake is the most important part of data analysis and processing. It is necessary to calculate the coordinate of the event and determine the energy magnitude of the event.

① **P-wave localization method in HSR earthquakes**. Since P-wave propagates the fastest among earthquake waves and the initial time of arrival is easy to identify, P-wave localization is adopted. It is assumed that the rock stratum is a uniform velocity model, P-wave propagation velocity is known, and monitoring stations are to be set up at least at four different locations.

In this chapter, the uniform velocity model is used to define the seismic center of the HSR. It is assumed that the rock stratum between the focus and each station is uniform, and then, P-wave propagation velocity v is a constant value. Let 0 be the focus, with the coordinate (x_0, y_0, z_0).

$T_i(i = 1, 2, \ldots, n)$ is the monitoring HSR station, and the coordinate of each HSR station is (x_i, y_i, z_i) $(i = 1, 2, \ldots, n)$;
$l_i(i = 1, 2, \ldots, n)$ is the distance from each HSR station to the focus;
$t_i(i = 1, 2, \ldots, n)$ is the moment when P-wave reaches each HSR station,
t_0 is the moment when the focus is generated.

The moment when P-wave arrives at each HSR station in the HSR earthquake is

$$t_i = \frac{l_i}{V} + t_0 \tag{2.14}$$

The distance from the HSR station to the focus in the earthquake is

$$l_i = \sqrt{(x_i - x_0)^2 + (y_i - y_0)^2 + (z_i - z_0)^2}, \tag{2.15}$$

where $t_i(i = 1, 2, \ldots, n)$, V, (x_i, y_i, z_i) $(i = 1, 2, \ldots, n)$ are known quantities. While the focus location of the microearthquake event is (x_0, y_0, z_0), and the moment of the focus generation t_0 are unknown quantities, which need to be solved.

Let $\bar{t}$ be the average moment at which P-wave arrives at each HSR station and $\bar{l}$ be the average distance from each HSR station to the focus, then

$$\bar{t} = \frac{1}{n}\sum_{i=1}^{n} t_i. \tag{2.16}$$

Constitute the least-squares function $\min f_k = \sum_{i=1}^{n} \left(t_i - \bar{t}\right)^2$ and solve its least-squares solution to obtain the focus location (x_0, y_0, z_0) and the moment of the focus generation t_0.

② **Ray method of the HSR earthquake wave**. In the process of microearthquake monitoring, some microearthquake may hit fewer than four HSR stations. They can be located using the ray method of the earthquake wave or the intersection of the propagating direction of P-wave rays. When there is only one station hit, the ray method of the earthquake wave can be used to get the focus location. The functional relationships are as follows.

The moment when P-wave arrives at each HSR station in the HSR earthquake is

$$t_p = \frac{V_s \Delta t}{V_p - V_s}. \tag{2.17}$$

The time difference between P-wave and S-wave in the HSR earthquake is

$$\Delta t = t_s - t_p = T_s - T_p. \tag{2.18}$$

The distance from the focus to the HSR station in the earthquake is

$$L = V_p t_p, \tag{2.19}$$

where

T_0 is the origin moment of the focus;
L is the distance from the focus to the HSR station;
T_p is the moment when P-wave arrives at the HSR station;
V_p is the propagation velocity;
t_s is the travel time;
Δt is the time difference between P-wave and S-wave.

The distance from the focus to the monitored HSR station is calculated. After that, according to the earthquake waves received by the station, the initial P-wave amplitudes in the east–west, north–south and vertical directions are measured, and the azimuth of the focus α can be found according to the formula. Then

$$\tan \alpha = \frac{A_{\mu E}}{A_{\mu N}}, \tag{2.20}$$

where $A_{\mu E} = \frac{\alpha_{\mu E}}{V_E}$, and $A_{\mu N} = \frac{\alpha_{\mu N}}{V_N}$.

The azimuth of the focus in the HSR earthquake is

$$\alpha = \arctan\left|\frac{A_{\mu E}}{A_{\mu N}}\right|, \tag{2.21}$$

where

$A_{\mu E}$ is the displacement of P-wave in the east–west direction (unit: μm);

$A_{\mu N}$ is the displacement of P-wave in the north–south direction;
$\alpha_{\mu E}$ is the amplitude of the initial P-wave in the east–west direction;
$\alpha_{\mu N}$ is the amplitude of the initial P-wave in the north–south direction;
V_E and V_N are the magnifications in the east–west and north–south directions, respectively.

2.5.3 *Earthquake Warning for HSR*

China is an earthquake-prone country, and earthquakes pose a serious threat to its HSR operation. By drawing on the experience of foreign HSR earthquake early warning, an efficient, reliable and feasible HSR earthquake warning system is established, which is a key to the construction of HSR in China's seismic zones.

A. **Main components of the earthquake warning system for HSR**. According to the current situation, the HSR earthquake warning system in China includes:

    Component I: Based on earthquake sensors, realize the interface design for earthquake information acquisition.
    Component II: Based on the 802.11-WIFI wireless transmitting module, realize real-time transmission of sensor data.
    Component III: Design the data fusion processing center of the upper computer based on the Visual Studio platform.
    Component IV: On the basis of traditional earthquake warning and judgment, comprehensively considering the impact of earthquakes on train operation, an assessment algorithm for HSR operation safety based on earthquake intensity is designed.
    Component V: Based on mobile communication base stations, push assessment information about HSR operation safety.

B. **The design process of HSR earthquake warning system**. The HSR warning system based on earthquake detection can be divided into four parts: threshold discrimination, early warning algorithm, system structure and integration testing.

    (1) **Definition of the HSR earthquake warning threshold**. When the intensity of the earthquake reaches the set threshold, the warning system issues an alarm. Indicators of the earthquake intensity include intensity and ground motion acceleration. The alarm threshold of the HSR earthquake is generally expressed by the gal value. The HSR earthquake alarm threshold $f(\text{EAT})$ is calculated by Eq. (2.7) (Table 2.6).

    The two main indicators that characterize the intensity of earthquakes are magnitude and intensity. Magnitude is a measure of the force of an earthquake, reflecting the difference in energy released by different earthquakes. It is determined by the vibration amplitude of the ground motion recorded by seismographs. Intensity is the degree of damage caused by earthquakes at different locations, divided into 12 levels.

Table 2.6 Definition of earthquake earning threshold

Level	Acceleration of ground motion/gal	Earthquake magnitude	Earthquake intensity	Operation regulations
1	[0, 40)	[1, 3)	[1, 4)	Decelerate by 0–30%
2	[40, 60)	[3, 4.5)	[4, 7)	Decelerate by 40–60%
3	[60, 90)	[4.5, 6)	[7, 9)	Decelerate by 60–80%
4	[90, 120)	[6, 8)	[9, 12)	Stop
5	> 120	> 8	> 12	Stop

(2) **Algorithm for HSR earthquake warning**. Parameters such as the distance from the epicenter to the railway line and the magnitude of the earthquake are estimated based on detected P-waves. Magnitude and the epicenter distance are used to determine whether the trains need to be controlled and the scope of the traffic control. Now, we use the earthquake formula (2.5) proposed by scholar K. C. Peng in 1985 for North China, which is $\lg A = -0.474 + 0.613M - 0.873 \lg R - 0.00206R$. Therefore, an alarm is required when a train is within the damage radius.

(3) **Process of the HSR earthquake system**. The structure of HSR earthquake data information collection, evaluation, alarm and control devices is shown in Fig. 2.17, including earthquake information acquisition module, storage module, communication module, monitoring host, alarm release module and the control module. The early warning process of the HSR earthquake system is shown in Fig. 2.18.

The specific process of HSR earthquake warning is as follows:

Step 1: Obtain parameter information: The surface motion information of the location with the acquisition module.
Step 2: The computer calculates earthquake parameters and quickly estimates the affected area based on the latest surface motion parameter information currently collected.
Step 3: Read the preliminary set threshold of the HSR earthquake intensity.
Step 4: Determine whether the set earthquake intensity threshold is exceeded.
Step 5: If the earthquake intensity information of the line does not exceed the safety threshold, return to Step 1. If the control module recognizes that the earthquake intensity exceeds the safety threshold, the control module activates the warning device and the process enters Step 6.
Step 6: The warning device is activated and warning starts.
Step 7: When receiving the warning signal from the early warning device, the spatial coordinates of the earthquake information acquisition module will be immediately obtained to locate the affected line.

Fig. 2.18 HSR earthquake warning process

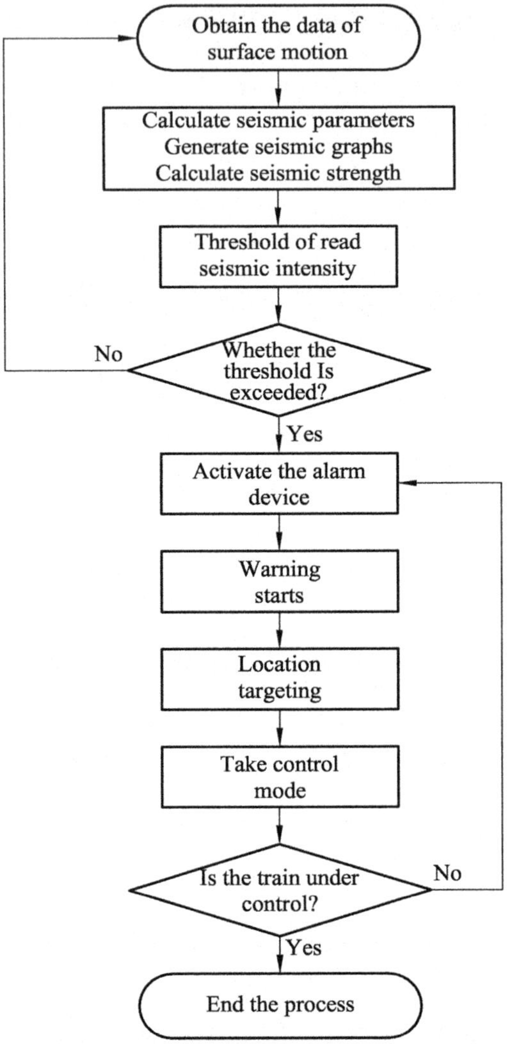

Step 8: The control module starts the vehicle control mode and the module drives the relay to control the traction power.
Step 9: Judge whether the relay is successfully disconnected and whether the train is controlled. If it is not, return to Step 5 and issue an alarm. If the train is effectively controlled, enter Step 10 and exit the process.
Step 10: The train is controlled, exit the process.

(4) **Test of the HSR earthquake warning system**. The input of earthquake waves is an important basis for seismic analysis. The peak acceleration of the high-speed train derailment is not the same and varies with different earthquake waves.

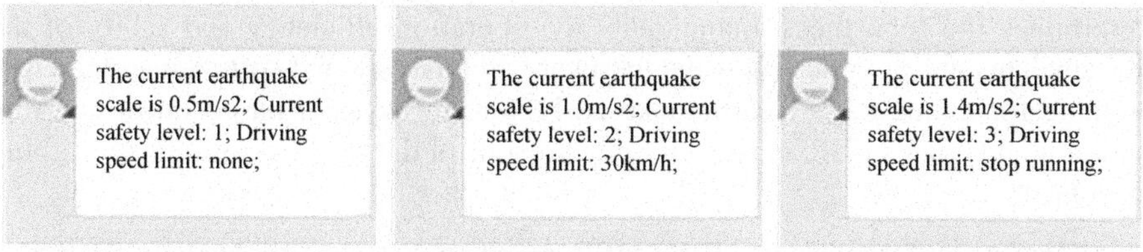

(a)Warning result of low-level earthquake (b) Warning result of low-level earthquake (c)Warning result of low-level earthquake

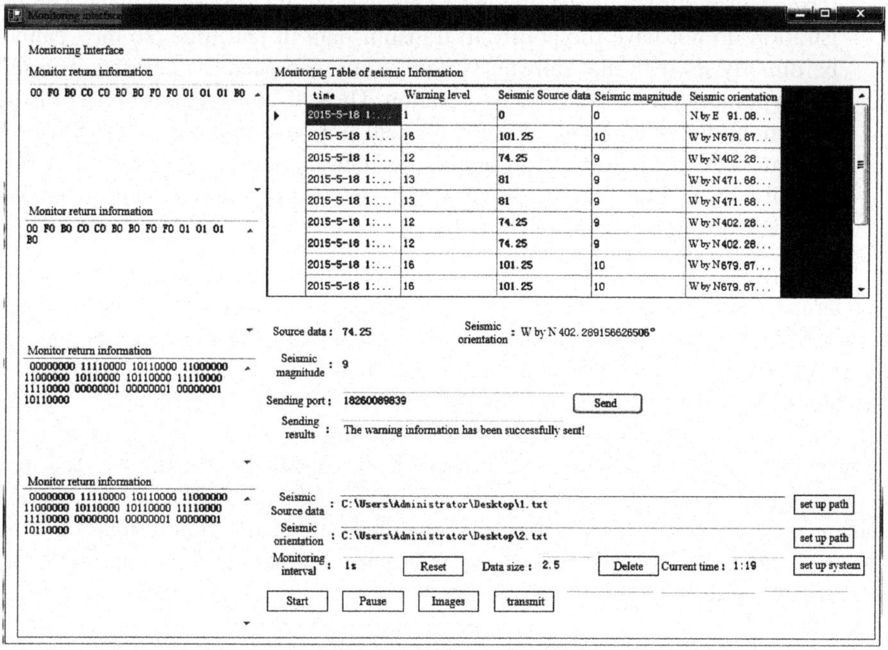

(d) Interface of HSR warning system

Fig. 2.19 Test results of HSR early warning system based on seismic grade detection

Based on the existing theory and program implementation, different control modes of the system under different earthquake levels are studied. In actual tests, there are three warning levels. The final test results obtained are shown in Fig. 2.19.

2.5.4 *Composition of the HSR Earthquake Warning System*

The HSR earthquake warning system is composed of data acquisition terminals, high-speed communication network, data processing center, alarm devices and control devices. The architecture of the HSR earthquake warning system affects and even determines the robustness, maintainability, operation efficiency and safety of the system. This chapter constructs the hardware architecture and software architecture suitable for China's HSR earthquake early warning system based on the current situation of China's HSR and development results of the world's earthquake warning systems.

(1) **Hardware framework of the HSR earthquake warning system**. The most important infrastructure in the HSR earthquake warning system is the network formed by multiple seismic stations. Some stations of China Earthquake Administration do not have the ability to transmit data in real time, so they cannot be directly used by the railway system. They can serve as peripheral monitoring points after real-time transformation. The self-built stations of the railway system, on the other hand, all meet the real-time requirements with their own communication networks. Combining the National Earthquake Networks with the railway self-built earthquake networks can improve the calculation accuracy and provide more response time.

The hardware of the HSR earthquake warning system includes data acquisition terminal equipment, monitoring host, communication facilities, data processing server, web server and database server. The data acquisition terminal equipment and the monitoring host are arranged at each traction substation. Data processing server and the database server are arranged at each railway section subcenter to facilitate the management and control of trains. The communication network connects the subcenter of the railway section with each station under its jurisdiction. The data acquisition terminal uses a strong-motion seismograph, which consists of a strong seismic recorder and a force balance sensor. The data acquisition terminal is with the most risk in the risk analysis of the earthquake warning system. During detection, the status monitoring module in the monitoring host sends a test command to the strong-motion recorder and then detects the waveform returned by the strong-motion recorder. If no waveform is returned from the specified event segment, or if the returned waveform is distorted, a false message will be generated. The recorder needs to be synchronized with the time server of the railway system. The recorder designed with an embedded system can easily realize clock synchronization by using the NTP service of the embedded Linux system (Fig. 2.20).

The communication facilities of the HSR earthquake warning system include the hub, twisted pair, optical terminal, fiber optic cable, etc. The network topology adopts a star network, which is simple and easy to control and manage. It can detect and isolate faults easily. The data processing server analyzes the arrival time of seismic phase of each station, estimates the earthquake magnitude, epicenter and damage degree based on the information such as the geographic location of each station or active fault in the database server, and generates Shake Map, etc.

Due to the control mode of the traction substation system, a monitoring host is required to drive the relay to control the traction electricity. It is advisable to use a stable and reliable industrial control computer as the monitoring host of the HSR earthquake warning system. Depending on the early warning mode, the task of the monitoring host may also include the automatic distribution of data packets of the

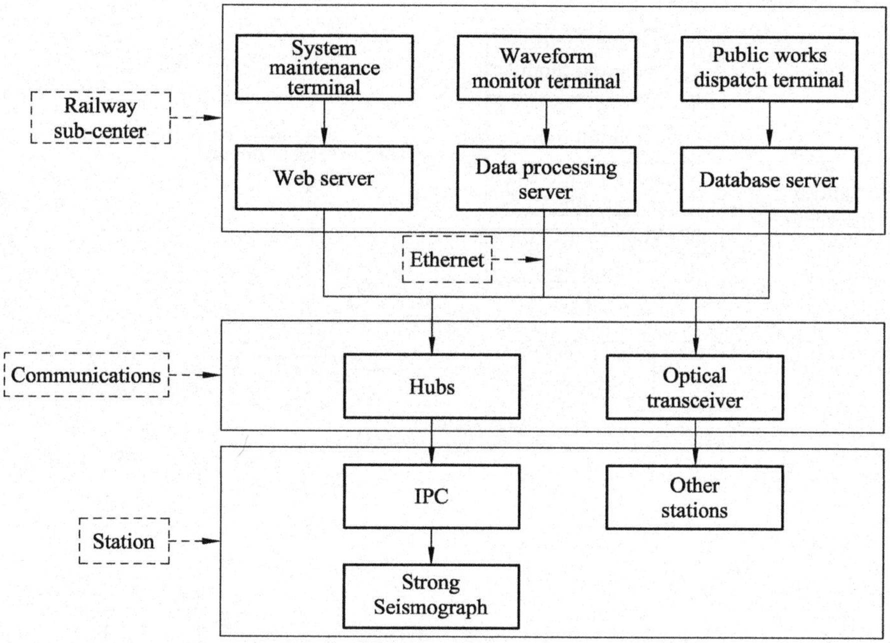

Fig. 2.20 Hardware architecture

strong-motion seismograph and the pickup of seismic wave phases, etc. If a single-station early warning mode is adopted, the monitoring host analyzes the earthquake data of this station and decides the measures to be taken.

(2) **Software framework for the HSR earthquake warning system**. The design objectives of the software framework of the HSR earthquake warning system are modularity, scalability, robustness and interconnectivity. A shared memory area is used in the software system to form a ring message queue for caching data and matching individual modules with different data processing speeds. The modules read their respective configuration files at startup and define the message queues they monitor and the message queues to which they output processing results. These files also define the operating parameters of the module, such as heartbeat interval, listening ports, filter parameters, etc. Each module of the HSR earthquake warning system is supposed to have a logging function to record the events that occur in each module and the corresponding time in detail. The granularity of the log output can be controlled by configuration files, with detailed information output during the development or testing phase and output the necessary abbreviated information after the deployment to the early warning system. The required software modules and interactions in the HSR earthquake warning system are shown in Fig. 2.21.

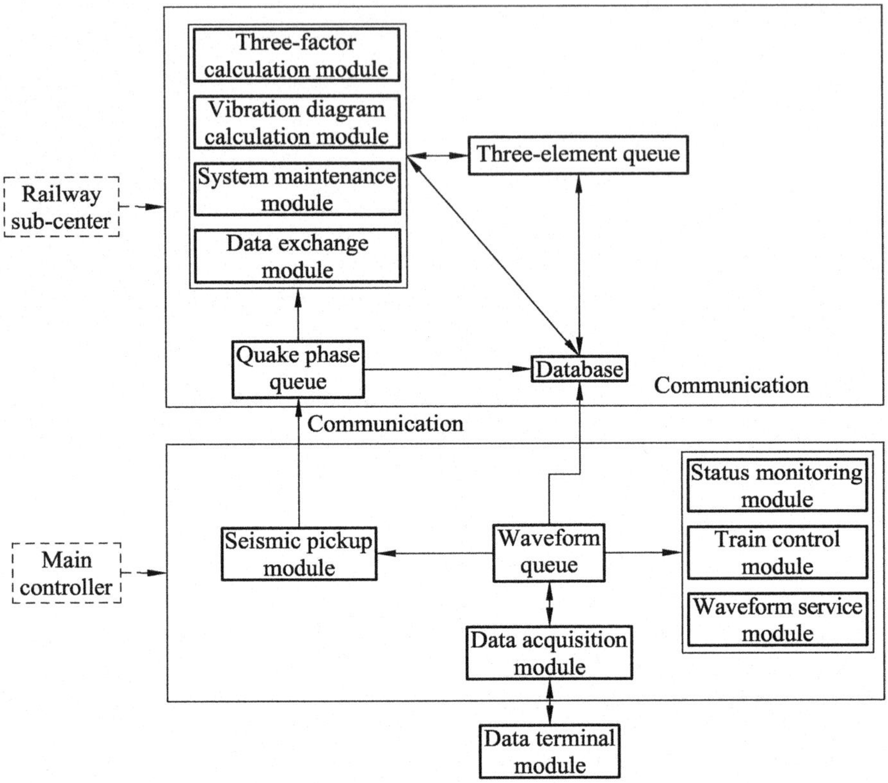

Fig. 2.21 Software architecture

The Open Service Gateway Initiative (OSGi) framework is an ideal choice if the strong earthquake recorder of the HSR earthquake warning system adopts the embedded system. OSGi is a service platform in Java language that supports the fully dynamic component pattern. Applications in the form of components (also called Bundles) can be installed, started, updated and uninstalled arbitrarily without the need to restart the program. Components can be started when needed to save resources. Different components can be combined into different complex functions. The Shake Map is a graphical representation of the ground motion caused by an earthquake, which can be used in the HSR earthquake warning system for a quick estimation of the affected area. This makes it possible to take train control measures only for trains in the area affected by the earthquake. After the earthquake, operation can be resumed by calculating the seismogram according to the complete station network records.

The HSR earthquake warning system can provide early warning information to the high-speed train control system in a timely manner and strive to take effective measures before the arrival of dangerous seismic waves to ensure traffic safety. This chapter studies the hardware and software modules required in the HSR earthquake warning system and their interrelationships and provides a good framework for the HSR earthquake warning system, which helps to ensure the robustness, scalability and safety of the early warning system.

2.6 Summary of This Chapter

Earthquake is the most threatening natural disaster to HSR operation safety. The current earthquake forecasting technology is still immature, so the development of the earthquake early warning technology is now an important measure to mitigate or avoid hazards to HSR. Considering domestic and international research on the HSR earthquake warning, this chapter compares the similarities and differences of HSR earthquake warning systems in various countries and establishes an early warning system suitable for China's specific situation with respect to the frequency or location of earthquakes in China. This chapter also innovatively proposes the definition of warning thresholds, which are defined quantitatively in terms of ground acceleration and earthquake intensity, and different control measures are taken for different warning levels, such as deceleration or stopping. In addition, this chapter gives a method for determining the threshold definition indicators, defines the whole process of the earthquake warning, completes the corresponding entity program that connects to the users and improves the warning program. Therefore, this warning system has certain applicability.

Bibliography

1. Park Y, Han S (2019) Development of disaster risk index for evaluating the natural disaster hazards of high-speed railroad facilities. Korean Soc Hazard Mitigation 19(3):1–9
2. Dahiya G, Asakura Y (2021) Exploring the performance of streaming-data-driven traffic state estimation method using complete trajectory data. Int J Intell Transp Syst Res 19(3):572–586
3. Ulak MB, Yazici A, Zhang Y (2020) Analyzing network-wide patterns of rail transit delays using Bayesian network learning. Transp Res Part C Emerg Technol 119:102749
4. Ministry of Railway Government of PRC, Asian Development Bank (2010) Railway emergency management system study. China Academy of Railway Sciences, Beijing
5. Hu Q, Zhang W, Zhang X (2014) Measurement theory and monitoring method for safe operation of high-speed railway. Science Press
6. Zhang H, Zhang Y, Xia D (2011) The design and development of the integrated disaster prevention and safety monitoring simulation system for passenger dedicated lines. China Railway Sci 32(1):136–140
7. Álvarez-SanJaime Ó, Cantos-Sanchez P, Moner-Colonques R (2020) Pricing and infrastructure fees in shaping cooperation in a model of high-speed rail and airline competition. Transp Res Part B Methodol 140(10):22–41
8. Sun H, Wang L, Dai X, Mu E, Liu J (2007) Study on the earthquake urgent automatic alarm system of high-speed railway. China Railway Sci 28(5):121–127
9. Zhang Y, Liu L, Liu M (2009) Optimization research on seismic array layout based on earthquake early warning system. J Shenyang Jianzhu Univ (Nat Sci) 25(1):1–5

10. Zhou S, Zheng J (2008) Earthquake countermeasures for high speed rail in Japan. J Railway Eng Soc 12(6):35–43
11. Li W (2004) Investigation report on disaster prevention and safety early warning system of French high-speed railway. Investigation Report of the Fourth Railway Survey and Design Institute, Wuhan
12. Yang J, Liu C (2006) Research on the derailment accident of the Joetsu Shinkansen in Japan. Chinese Railways 44(8):42–45
13. Misu Y et al (2008) Study on the strong wind estimation and train operation with wind observation and numerical analysis. In: The 20th symposium on wind engineering
14. Zhang X (2012) Research on seismic monitoring system of high-speed railway. Chinese Railways 50(8):51–53
15. Li L, Chen S, He Y (2008) Research on the earthquake response monitoring system for high-speed railway bridge. J Railway Eng Soc 12(6):169–175
16. Ma Q, Li S, Yu H, Song J (2013) Determination of emergency handling range of earthquake disaster prevention system for high-speed railway. J China Railway Soc 35(6):110–115
17. Nakamura Y (2004) UrEDAS, urgent earthquake detection and alarm system, now and future. In: 13th world conference on earthquake engineering. International Association for Earthquake Engineering, Vancouver
18. Wang Y (2012) Discussion on the interface scheme of high-speed railway signal system and earthquake monitoring system. Railway Signalling Commun Eng 9(3):16–18
19. Zhao J, Zhang Z (2009) Development, application and suggestions on earthquake early warning system. Geol Bullet China 28(4):456–462
20. Li S, Hong Z, Gao L (2011) An overview of earthquake early warning systems for high speed railways. World Earthq Eng 27(3):89–96
21. Ye K (2012) Research on earthquake early warning system for Beijing-Shanghai high-speed railway. Chengdu University of Technology

Chapter 3
A Lightning Warning System for HSR Operation Safety

Since the start of large-scale HSR operation, lightning strikes have been one of the main factors causing the failure of the traction power supply system and signal system. They not only damage the HSR equipment and cause losses of the train power, but also lead to transport interruptions. In some serious cases, lightning strikes may cause traffic accidents and casualties. For example, on July 23, 2011, the traction power supply catenary of the railway along Wenzhou South Railway Station in China was struck by lightning, resulting in a rear-end collision accident of the D301 and D3115 trains (see Fig. 3.1). Within 7 min after the accident, lightning strikes accumulated nearly 100 times, causing 40 deaths and 172 injuries, interrupting traffic for 32 h and 35 min and resulting in direct economic losses of about RMB 10 m.

Since there are many elevated bridges, complex climate types along railway lines and the use of integrated ground lines, China's HSR has a high probability of being struck by lightning and big risk of lightning strikes in operation. China is using a traditional railway lightning protection system today, which is not very efficient. In particular, because of the differences in geology, climate and structure of the power supply system, China can not apply lightning protection standards of other countries with developed HSR, such as Germany or Japan. Therefore, it is necessary to build a lightning protection system suitable for China's HSR system in light of its own conditions to guarantee HSR operation safety.

At present, the operating mileage of China's HSR ranks first in the world, but China's HSR lightning protection technology is developed for the general railway, which has low requirements for reliability of power supply. Also, general railways basically operate without viaducts. Therefore, the experience of general railway lightning protection may not be applicable to HSR. China's high-speed railways are mostly long-distance lines, such as Wuhan–Guangzhou line (HSR from Wuhan to Guangzhou, 1069 km), Beijing–Shanghai line (HSR from Beijing to Shanghai, 1318 km), which crosses multiple provinces. However, there are more thunderstorms in the south and fewer in the north. Areas are far apart and lightning activities vary

Q. Hu, *Natural Disaster Warning System for High-Speed Railway Safety Operation*, Advances in High-speed Rail Technology, https://doi.org/10.1007/978-981-99-7115-2_3

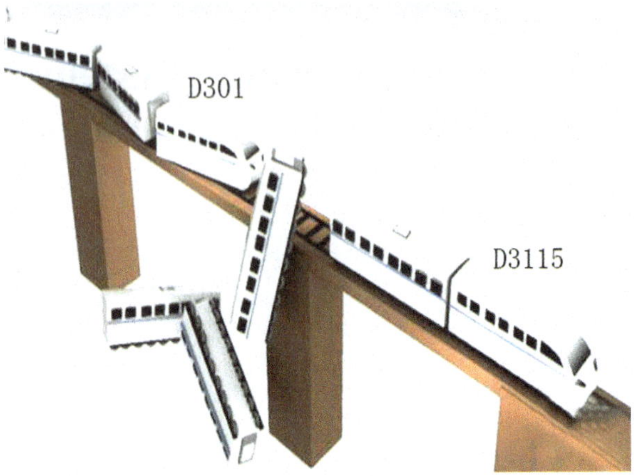

Fig. 3.1 Schematic diagram of the "July 23" accident

greatly across railway lines. These determine the particularity and complexity of China's HSR lightning disaster warning system.

3.1 Status Quo Analysis of the HSR Lightning Warning System

The HSR system is an automatic control system integrating various high technologies. It is composed of several subsystems, including the communication signal subsystem (CSS), truck side acoustic detection system (TADS), trouble of moving freight car detection system (TFDS), track hotbox detection system (THDS), truck performance detection system (TPDS), train coach running diagnosis system (TCDS), operation safety monitoring system (OSMS) and safety monitoring and control system for disaster prevention (SMCS). These systems are important factors to ensure HSR operation safety and prevent accidents and are important technical equipment to ensure the punctual, efficient, high-density and uninterrupted operation. The HSR system is like a complex neural network. Only when the communication signal transmits smoothly and the electronic equipment functions normally can the whole HSR system operate safely and efficiently, or it will fall into chaos. Therefore, the HSR lightning protection and early warning system is very important to the HSR system.

3.1.1 German HSR Lightning Warning System

The actual measurement of German HSR shows that there may be a lightning strike every 100 km of catenary in central European very year. Therefore, in the lightning

Fig. 3.2 German high-speed rail system

protection design of the catenary, direct lightning protection is not considered. Only lightning arresters are used to limit the induced lightning overvoltage. Due to the small number of lightning strikes, the use of automatic reclosing means can fully ensure reliable power supply (Fig. 3.2).

3.1.2 Japan's HSR Lightning Warning System

In the lightning protection design of Japan's electrified HSR, according to the frequency of lightning strikes and the importance of the line, there are three lightning protection levels in A, B and C areas, and corresponding lightning protection measures are stipulated, as given in Table 3.1.

Due to the low frequency of lightning strikes in Europe, lightning protection measures of European HSR are very simple, and there is no effective lightning protection warning system for HSR; and Japan's HSR lightning protection system is relatively complex and mature, worthy of reference.

Table 3.1 Lightning protection and early warning measures for overhead catenary of HSR in Japan

Area	Lightning protection measure	Overhead ground wire	Set position of the arrester
Area A	Lightning protection should be provided for important lines subject to serious lightning damage	Set up lightning lines all the way	Traction substation outlets; catenary disconnecting switches on both sides; connection of overhead lines and cables; overhead line terminals
Area B	For important lines with serious potential lightning damage, necessary lightning protection should be provided for related sites and key equipment	Lightning lines are set up along the catenary in places of particular importance	Traction substation outlets; catenary disconnecting switches on both sides; connection of overhead lines and cables; overhead line terminals
Area C	Outside areas A and B	–	Traction substation outlets; catenary disconnecting switches on both sides; connection of overhead lines and cables

3.1.3 China's HSR Lightning Warning System

The communication signal system adopted by China's HSR is the most advanced microelectronic equipment in the world. Many sensors are distributed along or near railway lines to collect traffic information. HSR tracks are also channels to transmit traffic information. These electronic devices are the core of lightning protection. As the traditional HSR communication signal system equipment uses separated electronic components, the damage energy of the separated transistor PN junction is 1 J, while the damage energy of the current large-scale integrated circuit chip is only 10–8 J. China's Ministry of Transport has taken lightning protection measures for the HSR communication signal system, but the HSR is often struck by lightning. Therefore, it is necessary to have a deep understanding of the characteristics of the HSR signal system to do a better job in HSR lightning protection. The concept of "integrated lightning protection" in the HSR communication signal system is shown in Fig. 3.3.

The HSR signaling system equipment relies on computer technologies, and more than 80% of the lightning damage to the HSR signaling system equipment is on these devices. For the HSR signaling system, lightning may enter the equipment in the following ways.

Way I: intrusion from the power line. In the HSR system, the power supply of the railway signal is generally from two inputs: one is dedicated to the HSR signal power supply (dedicated power) and the other is the power supply of the electrification section (traction power). The relationship between the two power supplies is active and standby with the dedicated power for active and the traction power for standby. Buried cables are used to transmit signals to the machine room, and the lightning in

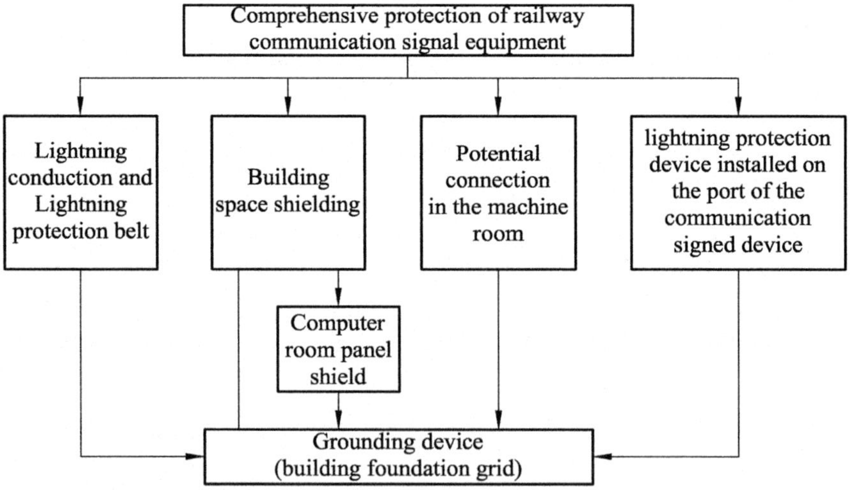

Fig. 3.3 Schematic diagram of lightning protection for HSR communication signal system

or near the power grid can be transmitted into the machine room through the power line.

Way II: intrusion by rail. All stations and grouping stations in the HSR system are equipped with signal equipment, most of which is installed along HSR lines. The information that displays the train dynamics and directs the train movement is transmitted on the steel rails, that is, the train is also the transmission line of blocked information. The HSR is composed of electrified sections. The track will not be directly hit by lightning under the shelter of the traction power supply network, but it may suffer secondary lightning strikes caused by direct lightning strike. The influence of the lightning electromagnetic pulse interference is inevitable, so the track in the HSR system is also an important way for lightning intrusion to signal equipment.

Way III: intrusion by the signal transmission line. The information from the intervals and stations in the HSR system is transmitted by data cables. The longest cable into the signal room is 10 km, and the shortest is more than 1.5 km. Although these cables are buried, due to the laying characteristics of the signal cables, they are subject to the risk of lightning electromagnetic pulse induction.

China's HSR is characterized by many elevated bridges and tunnels. The "buried earth cable (BEC)" and signal cables are buried in the same gully or near the bridge and tunnel section. There are two main tasks of the BEC: return part of the locomotive traction current and connect grounding subsystems of HSR to form a large HSR integrated grounding system.

Suppose there is a strike lightning in the distance, and the falling lightning point is point 0. When it is struck by lightning with current amplitude of i_O, the lightning electromagnetic field intensity H_O at point A from the lightning point 0 is shown in Fig. 3.4.

The lightning electromagnetic field strength at point 0 of the lightning fall is

$$f(H_o) = \frac{i_o}{2\pi s_a} H_O. \tag{3.1}$$

Direct lightning is not perpendicular to the earth generally, so the lightning current can be analyzed by vector. The horizontal component of the lightning strike forms

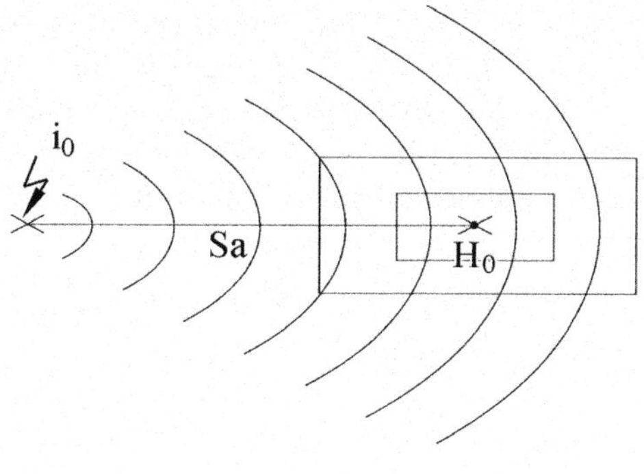

Fig. 3.4 Magnetic field value during nearby lightning strike

a closed loop with the BEC, the main reinforcement in the bridge pier, the ground network, the earth in Fig. 3.4. Therefore, the field strength of the lightning at a certain point s_a under a certain direct lightning current i_o can be calculated, and the lightning current on the signal sheath of the BEC and the induced lightning current on the cable core can be calculated.

3.1.4 Comparison of HSR Lightning Warning Systems in China and Abroad

China is a vast country and lightning phenomena are frequent, but the situation varies in different regions of the country. The main range of global thunderstorm activities is between 400 N and 400 S, and their intensity is relatively low in the areas north of the 400 N line. For example, the latitude of Heilongjiang Province of China is comparable to that of Germany. The latitude of the center of Germany is about 51° 4′ N, and that of Harbin is about 45° 4' N, so they are basically comparable in terms of thunderstorm activities. For Japan, its HSR has a more complete lightning protection system, but Japan's land area is small, which is close to the area of China's Shandong Peninsula, Liaodong Peninsula. Japan's lightning activities are not frequent, with only 25–30 lightning days per year. Therefore, China's thunderstorm activities are fundamentally different from those in Germany and Japan. In addition, China has a vast territory and great climate differences between different regions such as northeast, northwest and south China. Overall, foreign HSR lightning technologies cannot meet the actual needs of China, so a HSR lightning protection and early warning system suitable for its national conditions must be established.

3.2 Influence Mechanisms of Lightning on HSR Operation Safety

According to the interval of lightning action density and on the basis of comprehensive analysis of domestic and foreign research results, this chapter divides China's lightning risks into four levels: level one (serious areas), level two (generally serious areas), level three (minor areas) and level four (safe areas), see Table 3.2.

According to the classification of lightning risks in Table 3.2, the classification of lightning risks in China's provinces is given in Table 3.3.

According to Table 3.3, the current situations of lightning hazards in China are as follows:

Level 1 danger: Zhejiang, Sichuan and some other provinces are with serious lightning danger;

Level 2 danger: Guangdong, Jiangsu, Hainan and other nine areas are with general serious lightning danger;

Table 3.2 Lightning risks levels in China

Number	Danger level	Lightning danger	Lightning density interval/[Times/(km^2 year)]
1	Level 1	Serious area	50–70
2	Level 2	General serious area	30–50
3	Level 3	Minor area	10–30
4	Level 4	Safe area	0–10

Table 3.3 Division of China's thunder and lightning areas

Level	Density interval	Region	Lightning number	Maximum density
	Times/(km^2 year)		Number	Times/(km^2 year)
Severe areas	50–70	Zhejiang	480,076	76
		Sichuan	1,045,599	57.256
General serious area	30–50	Guangdong	890,135	48.25
		Jiangsu	545,433	40.25
		Hainan	154,481	38.75
		Anhui	476,831	35.75
		Fujian	411,407	35.5
		Jiangxi	600,154	35
		Shandong	189,081	32.5
		Guizhou	501,600	31.75
		Henan	256,751	30.67
Minor area	10–30	Hebei	423,712	29.997
		Hunan	399,485	29
		Shanxi	359,818	27.41
		Yunnan	610,940	27.21
		Guangxi	432,526	26.25
		Chongqing	218,773	23.58
		Shanghai	30,800	17.176
		Shaanxi	193,579	15.46
		Heilongjiang	371,687	15.44
		Beijing	44,583	13.61
		Jilin	129,387	9.25
Safe area	0–10	Liaoning	105,895	8.75
		Tianjin	24,008	8.6
		Ningxia	8,575	4.77

Level 3 danger: Hebei, Hunan, Yunnan and other 11 areas are with minor danger;
Level 4 danger: Liaoning, Tianjin and Ningxia are safe areas of lightning dangerous situation danger.

Therefore, different areas in China should adopt different control measures to reduce lightning strikes and ensure HSR operation safety.

From the perspective of the regions which China's HSR lines pass, the lines most vulnerable to lightning are Ningbo–Wenzhou line, Wenzhou–Fuzhou line, Shanghai–Hangzhou line, Chongqing–Lichuan line, Suining–Chongqing line and Dazhou–Chengdu line. Beijing–Shanghai line, Hefei–Bengbu line, Wuhan–Guangzhou line, Guangzhou–Shenzhen line, Fuzhou–Xiamen line, Hefei–Wuhan line, Hefei–Nanning line, Qingdao–Jinan line, Shanghai–Nanjing line and some other lines are with general lightning danger in China, while the Beijing–Shijiazhuang line, Wuhan–Yichang line, Yichang–Wanzhou line and Shijiazhuang–Taiyuan line are with slight lightning danger. Therefore, different lightning warning systems should be adopted for HSR lines in different regions.

3.2.1 Mechanism of the Lightning Impact on HSR Signal System

At present, there are various ways of lightning hazards to high-speed railways. Through comprehensive analysis, this chapter summarizes the forms of lightning hazard to high-speed railway in China into six types.

Hazard way 1. Lightning damages the HSR transmission and substation system. HSR substation is the hub of the HSR power supply. Once suffered a lightning strike, power outages will spread widely, affecting HSR operation.

Hazard way 2. Lightning damages the HSR signal system. The track is a good conductor of the lightning, and the signal equipment connected to the HSR track is vulnerable to lightning hazards.

Hazard way 3. Lightning damages the HSR communication system. Lightning damages the communication signal and network communication equipment through conduction and induction. By invasion to computer communication lines, lightning damages the HSR wireless communication equipment.

Hazard way 4. Lightning damages the catenary system of the electrified HSR. The damage to pillars and insulators in the HSR catenary system will cause permanent grounding fault in the traction system, forcing the interruption of power supply.

Hazard way 5. Lightning damages other weak HSR electrical systems, such as ticketing system, equipment monitoring system, automatic fire alarm system, etc.

Hazard way 6. Lightning affects the plateau line. Because the HSR operating environment is generally extremely harsh, the impact in areas of high altitudes and strong lightning environments on the HSR electrical products is significant.

Therefore, among these six hazards, damage to the HSR signal system appears most frequently, which is an important part of the HSR lightning protection technology research.

3.2.1.1 Vulnerability of the HSR Signal System in Lightning Environment

The HSR signal system is vulnerable to lightning strikes, which is related to the characteristics of the system itself. The HSR signal system is huge, and the equipment is widely distributed. Among them, the annunciator, turnout and other devices are connected to the track and other components for tens of kilometers. Moreover, the HSR signal system is directly exposed to the space atmosphere, and it is easy to be affected by lightning strikes.

(1) **Vulnerability of the signal building at HSR stations**. The signal tower in the HSR station is in the place where HSR signals are concentrated. The HSR signal system is required to be reliable, and the withstanding voltage of the HSR signal equipment is low. Compared with ordinary buildings affected by lightning strikes, the HSR signal building is vulnerable to lightning strikes mainly for the following reasons.

Geographical location factors of the signal tower at the station: most of China's HSR grouping stations and stations are far away from the city center. The HSR signal tower is mostly located in the open area of the high-speed railway station yard. Such an environment is prone to lightning strikes, and there are tall objects near the signal tower that are easy to cause lightning.

Characteristics of the signal tower at the station: First, the HSR signal tower is usually tall, which is vulnerable to lightning because the lightning is easy to strike protruding objects. Secondly, different from general buildings, it is the place where the signal systems of HSR stations are concentrated, and there are many microelectronic devices, providing a variety of ways for lightning intrusion. Finally, most of the signal devices in these signal towers are weak current equipment with low working voltage and low anti-electromagnetic interference and anti-overvoltage capabilities, which are easily affected by lightning induction and interfered with the normal operation of the equipment.

(2) **Vulnerability of the HSR track circuit**. The HSR track circuit is a good conductor to receive direct and induced lightning. Because the track is exposed to various terrains and experiences various climatic environments from cold temperate to subtropical, the thunderstorm situation is also very complex. In particular, the open area is more susceptible to lightning due to the higher frequency of thunderstorms, so the related HSR signal equipment is more prone to lightning threats.

3.2.1.2 Variability of Lightning Activities

The intensity of lightning activities varies from region to region, with some areas being strong and others weak. China is a vast country, with its HSR running through. The HSR lines experience various climatic and topographical conditions, so the intensity of lightning activities varies. China still expresses the intensity of lightning activities with the annual average lightning days today. Therefore, China generally has four types of regions based on lightning intensity.

Region I. Intensity of lightning activities in northwest China. The average annual thunderstorm days in northwest China are generally fewer than 20 days. Among them, the place with the weakest lightning in China is Xinjiang, with the least average annual thunderstorm days and the shortest thunderstorm season;

Region II. Intensity of lightning activities in the areas north of the Yangtze River.

Region III. Intensity of lightning activities in the areas south of the Yangtze River. The annual average thunderstorm days in the area south of the Yangtze River (north of 23° N) are generally 40–80 days;

Region IV. Intensity of lightning activities in southwest China. The annual average thunderstorm days in the areas south of 23° N in southwest China are more than 80 days.

Since the lightning activities in China are mainly concentrated in June to August, with the most frequent lightning activities in July. Therefore, the laws of lightning activities in China can be summarized as follows.

Law I. Lightning activities in China are more in the south than in the north. The further south, the closer to the equator and the tropics, the stronger the lightning activity; the further north, that is, the lower the temperature and the less rainfall, the weaker the lightning activity.

Law II. Lightning activities in China are more in the mountains than in the plains. In addition, the rainfall in China's mountains is generally more than that in the plains, so there are more lightning activities in the mountains than in the plains.

Law III. Lightning activities in China are more in the inland than in coastal areas. When other conditions are the same, lightning activities in areas near the sea or near large rivers are weaker than those in other areas.

The annual average thunderstorm days are a necessary indicator for lightning impact analysis and an important basis for lightning protection design. The statistical data of the average annual thunderstorm days around China are given in Table 3.4. Where there are more thunderstorm days in a year, the HSR signal system is more likely to be affected by lightning. Therefore, lightning protection there should be strengthened. Lightning protection level should be lifted in areas with more lightning activities.

Table 3.4 Average annual thunderstorm days in different parts of China

Region	Annual average thunderstorm days	Region	Annual average thunderstorm days
Shanghai	35	Xi'an	20
Beijing	40	Chongqing	40
Nanjing	38	Nanchang	60
Tianjin	30	Changsha	50
Guangzhou	90	Fuzhou	60
Harbin	80	Lanzhou	25
Shenyang	33	Taiyuan	40

3.2.2 Vulnerability of the HSR Traction Power Supply System in Lightning Environments

The HSR traction power supply system consists of substations and traction networks, as shown in Fig. 3.5. The voltage at the incoming side of the HSR substation is generally 220 kV, and the lightning protection at substations is consistent with that of the electric power system. The experience of the power system operation shows that the primary cause of the tripping of the high-voltage transmission network throughout the country is lightning, and the proportion of the tripping is 40–70%. The HSR traction network is composed of low-voltage power lines. The whole line is equipped with lightning conductors, and about 80% of the mare set up on viaducts. Therefore, the level of lightning resistance is even lower, very vulnerable to lightning strikes and failures. According to statistics, after the operation of Wuhan–Guangzhou, Beijing–Guangzhou, Hangzhou–Ningbo and other high-speed railways started, there were many serious situations in which trains were delayed severely and passengers were trapped for a long time due to the failure of the traction network caused by lightning strikes.

The lightning protection design of the catenary in China's electrified railway catenary is mainly based on the "Code for Design of Railway Traction Power Supply" (TB10009-2005) and the "Interim Provisions for Railway Lightning Protection, Electromagnetic Compatibility and Grounding Engineering Technology" (TJS [2007] No. 39). According to the number of lightning days, areas are divided into four levels, see Table 3.5.

The frequency of lightning strikes on the HSR catenary is related to the annual average thunderstorm days T_d. If T_d is larger, the number of lightning strikes per year per 1 square kilometer of land is also larger. According to the recommendation of the International Conference on Large Power Grids: when the side limit of the HSR catenary is 3 m, the average height of the carrier cable from the rail surface is 7 m.

① The number of lightning strikes on the catenary of single-track HSR:

$$N = 0.122T_d^{1.3}. \tag{3.2}$$

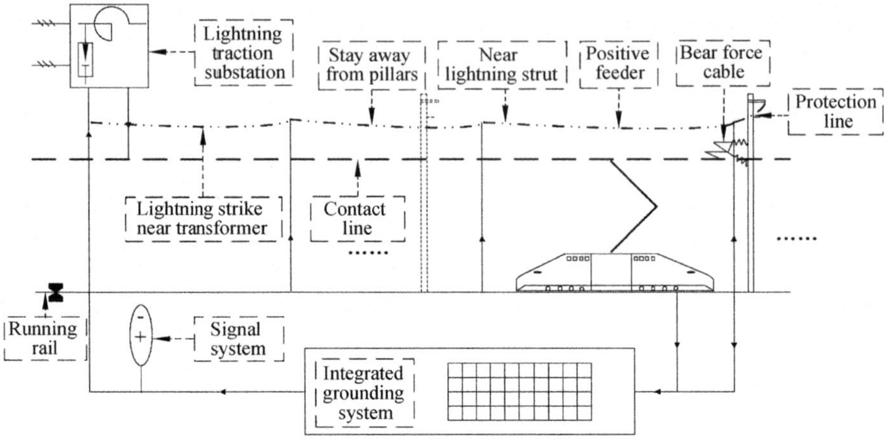

Fig. 3.5 HSR traction power supply system

Table 3.5 Lightning levels according to annual mean lightning days/day

Number	Level of lightning	Lightning area	Annual mean lightning days/day
1	Level 1	Fewer thunderstorm region	< 20
2	Level 2	More thunderstorm region	20–40
3	Level 3	High thunderstorm region	40–60
4	Level 4	Strong thunderstorm region	> 60

② The number of lightning strikes on the catenary of double-track HSR:

$$N = 0.244T_d^{1.3}, \tag{3.3}$$

where N is the frequency of lightning strikes on the catenary.

3.2.3 *Definition of Parameters for Lightning Hazard Warning*

Lightning parameters are used to describe the characteristics of lightning discharges, which are an important basis for the analysis of lightning impact. The occurrence of lightning is related to many factors such as topographic conditions, seasonal conditions and meteorology. For example, lightning activities in the ocean are greater than those on land, and those in mountainous areas are stronger than in the plain. Moreover, the occurrence of lightning is highly random, so the parameters describing the characteristics of lightning discharges are derived from statistical laws. In this chapter, the statistical data of lightning discharge parameters are obtained by comprehensive analysis based on the data obtained from long-term observations of typical thunderstorm areas in various countries.

(1) **Annual average thunderstorm days**. Thunderstorm days refer to days on which thunderstorms occur. As long as one or more thunderstorms are heard in a day, it is a thunderstorm day, regardless of the number and the duration of the thunderstorm. Thunderstorm days cannot reflect the actual number or the duration of thunderstorms in a day. The number of thunderstorm days varies greatly from year to year, so annual average thunderstorm days should be adopted. Areas with more annual average thunderstorm days are more likely to suffer lightning strikes. China's voltage protection regulations stipulate that

    Fewer thunderstorm areas: areas with fewer than 20 thunderstorm days a year;
    More thunderstorm areas: areas with an average of 20–40 thunderstorm days a year;
    High thunderstorm areas: areas with an average of 40–60 thunderstorm days a year;
    Strong thunderstorm areas: areas with over 60 thunderstorm days a year.

(2) **Annual lightning strike rate**. The annual lightning strike rate refers to the number of lightning strikes on a building or other object per unit area ~~per~~ year. The specific value is related to the equivalent area of the building, the current thunderstorm days and the ground conditions of the building. According to GB 50057-94 "code for the design of the building lightning protection," the estimated number of annual lightning strikes on the building can be calculated as follows:

$$N = kN_g A_g, \tag{3.4}$$

where

N is the estimated number of lightning strikes to the building, times/a;
k is correction coefficient for the number of lightning strikes, generally taken as 2;
A_g is the equivalent area of a building with the same number of lightning strikes, km^2.

(3) **Ground flash density**. From the perspective of lightning protection, to specifically describe the frequency of thunderstorm activities and express the frequency of thunderclouds and ground discharges, the most important parameter to determine the impact of ground lightning on buildings is the ground flash density, i.e., the number of lightning strikes per square kilometer on each lightning day, calculated as

$$N_g = 0.024T_d, \tag{3.5}$$

where

N_g is the average annual density of lightning strikes on the earth in the area where the building is located, times/$\left(\text{km}^2\ \text{a}\right)$;
T_d is the average number of annual lightning days in the area.

Equation (3.5) shows that the greater the average annual number of thunderstorm days, the greater the density of lightning strikes on the earth.

3.3 Lightning Warning System for HSR Operation Safety

According to the disaster warning information released by the meteorological department and the lightning information along HSR lined in the immediate future, if the possible loss of traction power supply system caused by lightning disaster is assessed before the outbreak of lightning disaster, it is possible to predict the failure of the traction power supply system caused by lightning strike, and to provide early warning for the HSR safety operation to improve the reliability of HSR operation safety in thunderstorm weather.

3.3.1 Early Warning Methods for Lightning Hazards to HSR

The international lightning protection standard IEC 62305 takes the lead in introducing risk assessment to the lightning protection of buildings, and proposes the risk of lightning damage to high-speed railways. The calculation function is

$$R = \text{NPL}, \tag{3.6}$$

where

R is the risk of lightning damage to high-speed railway;
N is the number of dangerous events in the HSR traction network;
P is the probability of damage;
L is the damage caused by the risk of lightning damage to HSR.

3.3.1.1 Risk Sources of Lightning Damage

In the past, the parameters characterizing lightning risk sources in China's design specifications of the HSR traction network were mainly based on meteorological thunderstorm days, but cannot accurately describe the lightning ground flash density and lightning current intensity. They could hardly quantitatively reflect the risk of lightning damage in each section in the traction network. At present, the power industry has built a lightning monitoring network covering 33 provinces (municipalities, autonomous regions and special administrative regions) in China expect Taiwan

Province, including about 500 lightning detection stations. After years of operation, 1012 bytes of lightning monitoring data have been accumulated. Detailed, accurate and true lightning density and intensity parameters can be obtained through statistical analysis of the data.

The span of China's HSR is very large. In order to obtain learn about the distribution of lightning hazard risk sources in high-speed railway corridor, the "line corridor grid method" can be used to divide the lines into several grids at equal distance. Considering the action scope of ground induced lightning, the positioning error of lightning monitoring network, and the actual needs of the project, the width of the transverse section along the line of the grid is selected as 5 km on the left and right side of the line. The frequency of risk source can be expressed by its density, that is, ground flash density (unit: times/$(\text{km}^2\ \text{a})$. The occurrence intensity can be expressed by the amplitude of lightning current.

3.3.1.2 Types of Lightning Damage

The 50% lightning impulse discharge voltage of high-speed railway traction network insulator is about 400 kV, so there are two types of lightning damage that may cause the traction network to trip, as shown in Fig. 3.6.

(1) **The damage mode of direct lightning strikes**. Overvoltage caused by direct lightning strikes on traction network of high-speed railway:

① Lightning strikes on *T* line (electric energy transmission line)/messenger wire, raising the potential of *T* line, exceeding the insulation level and causing flashover of *T* line insulator;

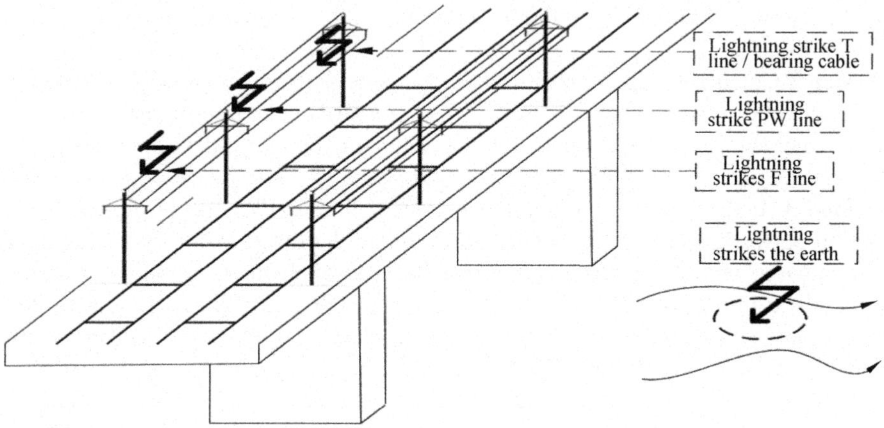

Fig. 3.6 Schematic diagram of lightning damage types of HSR traction network

② Lightning strikes on F line (feeder line, opposite phase with the T line), raising the potential of F line, exceeding the insulation level, causing flashover of line F insulator;

③ Lightning strikes on PW line (neutral line), and the lightning current enters the ground through the pillar and PW line, causing bigger potential of lightning strikes to the ground, exceeding the insulation level, and causing flashover of T line or F line insulator.

(2) **The damage mode of lightning induction**. The induced overvoltage is caused by lightning strikes on the ground or objects nearby, including the ground, mountains, trees, buildings, transmission lines, etc. Due to electromagnetic induction, the induced voltage on the conductor exceeds the insulation level, causing T line or F line insulator flashover.

3.3.1.3 Probability of Lightning Damage

The probability of the lightning damage to the HSR traction network is directly related to its lightning resistance level. It is also necessary to compare the overvoltage generated by the lightning strike with the insulation withstand voltage. The calculation process of the HSR lightning damage probability is shown in Fig. 3.7.

Calculation steps for the probability of damage caused by lightning strikes on HSR are as follows:

Step 1: Along the vertical line direction (i.e., Fig. 3.8 direction), within the distance from 0 to S (S should be greater than the maximum lightning current taken to calculate the farthest distance to make the traction network insulation flash, unit: m), lightning $F_i(x_i, I_i)$ is generated for n times at random, where x_i is the position coordinate of lightning F_i in the z direction, subject to uniform distribution; I_i is the current amplitude of lightning F_i, which follows the cumulative probability distribution.

Step 2: Calculate the geometric strike distance according to the electrical geometry model and judge whether lightning $F_i(x_i, I_i)$ strikes the traction network or the ground according to the relationship between the lightning strike position and the exposed arc.

Step 3: If the lightning strikes the F line, T line or PW line of the HSR traction network, the EMTP is called to calculate the direct strike overvoltage and compared with the corresponding insulation level. If the overvoltage is greater than the insulation level, the lightning strike will cause the insulation to flash over and cause the traction network to trip, and the number of direct strike trips will be added to 1. Conversely, the next lightning $F_{i+1}(x_{i+1}, I_{i+1})$ will be tripped. If the lightning strikes the earth, the electromagnetic field values are called to calculate the induced overvoltage and compared with the corresponding insulation level. If the overvoltage is greater than the insulation level, the lightning will make the insulation flashover, resulting in traction network tripping, the number of induction trips plus 1. On the contrary, skip the next lightning $F_{i+1}(x_{i+1}, I_{i+1})$.

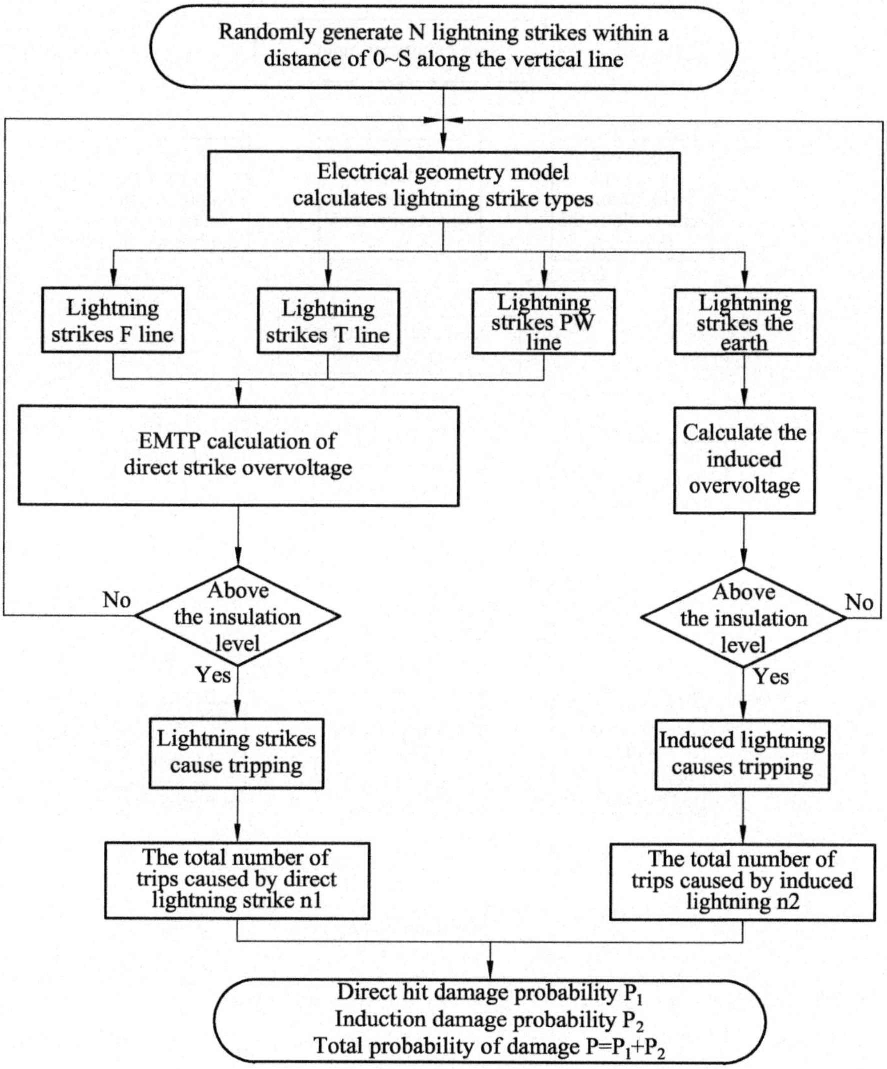

Fig. 3.7 Calculation process of lightning damage probability to traction network

Step 4: After the calculation of a lightning cycle, the total number of direct strike trips n_1 and induction trips n_2 will be calculated.

For each side of the HSR traction network within a distance of S, the probability of direct strike damage is $P_1 = \frac{n_1}{n}$, and the probability of induction damage is $P_2 = \frac{n_2}{n}$. The damage probability of total lightning is $P = P_1 + P_2 = \frac{n_1+n_2}{n}$.

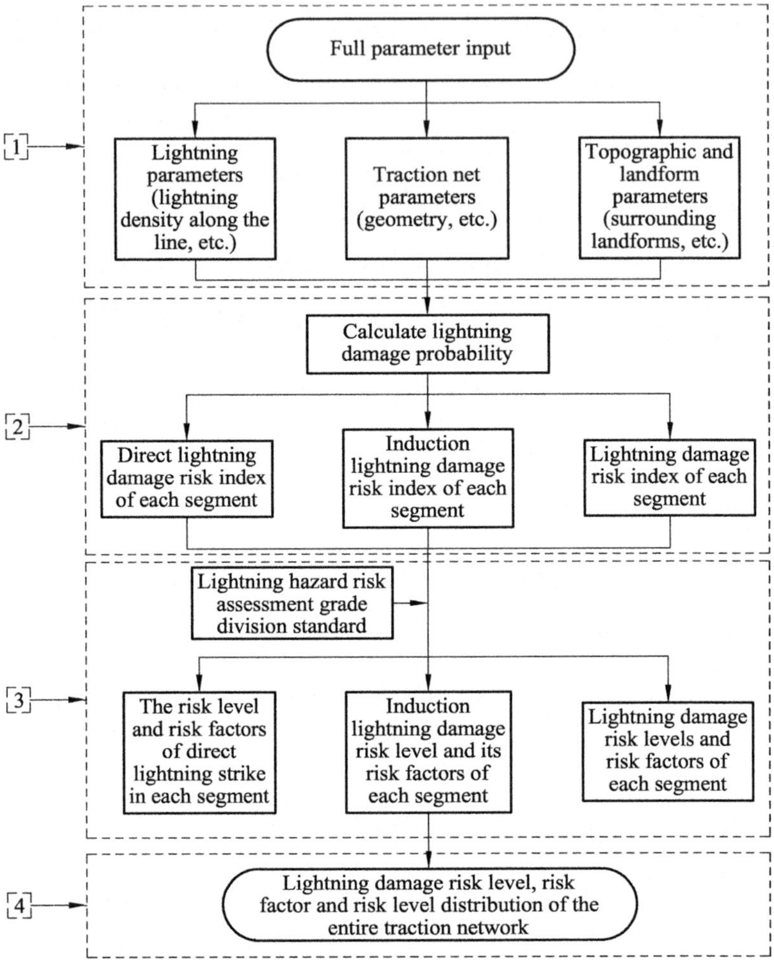

Fig. 3.8 Flowchart of traction network lightning hazard risk assessment

3.3.1.4 Risk Indicators for the Lightning Hazard

The risk indicator for the lightning hazard of the HSR traction network is the lightning trip rate of traction network. Let N refers to the number of lightning risk sources within the width of 2 s per 100 km unit length of the HSR traction network, and

$$N = \frac{s}{5} N_g, \tag{3.7}$$

where

N_g is the ground flash density;

S is the longest distance that the maximum lightning current can make the traction network insulation flashover, unit: m.

Then, the risk indicator for the lightning hazard to the HSR traction network:

(1) Risk indicator of direct lightning damage to the HSR system

$$R_1 = N \cdot P_1, \tag{3.8}$$

where

R_1 is the direct strike trip rate of the traction network in the HSR system.

(2) Risk indicator of the induction lightning damage in the HSR system

$$R_2 = N \cdot P_2, \tag{3.9}$$

where

R_2 is the induction trip rate of the traction network in the HSR system.

(3) Risk indicator of lightning damage to the HSR system

$$R = N \cdot P, \tag{3.10}$$

where

R is the lightning trip rate of the traction network in the HSR system.

3.3.1.5 Risk Assessment Level of the Lightning Hazard

To reflect the level of lightning hazard risk to the HSR traction network, the risk reference standard can be set and the risk level of the HSR traction network lightning damage can be determined according to the relationship between the risk assessment value and the reference standard. The risk reference standard can select the design value, actual operating value, average value of the assessment results or value specified by the operation management department of the lightning trip rate of the HSR traction network. The risk level of HSR traction network lightning is set according to risk assessment results. This chapter selects four-level risk classification method: low risk, medium risk, high risk and strong risk, represented by A, B, C and D, respectively, as given in Table 3.6, where R_s is the risk reference standard.

Table 3.6 Classification standard of lightning hazard risk assessment for HSR traction network

Risk indicator interval	$[0, 0.5\,R_S]$	$[0.5\,R_S, 1.0\,R_S]$	$[1.0\,R_s, 1.5\,R_S]$	$[1.5\,R_S, \infty]$
Risk level	A	B	C	D
	Low risk	Medium risk	High risk	Strong risk

3.3.1.6 Risk Assessment Process for the Lightning Hazard

The formation factors of HSR lightning disasters are complex. The invasion on the traction power supply system is instantaneous. If the loss of traction power supply system caused by lightning can be assessed according to the disaster warning information issued by the meteorological department and the forthcoming lightning disaster information along the HSR before the lightning occurs, it is possible to predict the failure of traction power supply system caused by lightning strike and to effectively warn to help ensure HSR operation safety. The risk assessment process of the lightning hazard to the HSR traction network is shown in Fig. 3.8. This chapter divides the assessment process into four parts.

Step 1: Input lightning parameters. Lightning hazard risk assessment parameters of the HSR traction network include lightning parameters, traction network parameters, topographic and geomorphic parameters, etc. Among them, the lightning parameters include the ground lightning density and the cumulative probability of lightning current amplitude at each section along the line, and the lightning parameters can be obtained from the monitoring data of the lightning monitoring network. Traction network parameters include 3D geometric structure parameters, line impedance, grounding resistance and other electrical parameters of each component, as well as insulation configuration. Traction network parameters are provided by the design or operation unit. The topographic and geomorphic parameters include the topography along the line, height of the viaduct, surrounding topography and buildings, etc. The topographic and geomorphic parameters are provided by the design unites or operation units, or obtained from the 3D geographic information system.

Step 2: Calculate risk indicators of the lightning damage. According to the input parameters, the calculation process of lightning damage probability of the HSR traction network is to calculate the direct damage probability, inductive damage probability and total lightning damage probability section by section. The direct lightning hazard risk indicators, inductive lightning hazard risk indicators and lightning hazard risk indicators of each section are obtained.

Step 3: Evaluate the risk level of the lightning damage. Set the classification standard of lightning hazard risk assessment levels as required, and pull the obtained risk indicators into the standard to obtain the direct lightning hazard risk level, inductive lightning hazard risk level, lightning hazard risk level of each section and the risk factors that determine the risk level.

Step 4: Determine the lightning hazard risk of the whole HSR traction network. Through the weighted average risk assessment results of each section, the risk level of the lightning damage and risk level distribution across the HSR line are obtained, and the main risk factors in the whole line are sorted out.

Following the above steps, the risk of the lightning damage to HSR traction network in different regions and at different times can be assessed comprehensively. The assessed level reflects the indicator of the lightning protection performance of the HSR traction network.

There is an interactive process between the risk assessment of the HSR lightning hazard and the operation status of traction power supply system. This chapter constructs the risk assessment and early warning system of the HSR lightning hazard, as shown in Fig. 3.9. The meteorological warning information includes the probability, intensity, density and falling area of the lightning. The risk assessment model of the HSR lightning disaster evaluates the possible impact of lightning on the operation of traction power supply system according to the meteorological warning information, determines the dangerous areas where the traction power supply system is susceptible to lightning and evaluates the degree of lightning impact on traction power supply equipment such as traction substations and contact networks along the line according to the information provided by the meteorological warning system, such as the probability, intensity and falling area of the lightning. The equipment with high risk of lightning strikes is identified to provide guidance for daily maintenance, disaster disposal and fault recovery of the traction power supply sector.

3.3.2 *Lightning Hazard Warning System for HSR Operation Safety*

Based on the analysis of the mechanism of the lightning disaster impact on the HSR, the HSR lightning disaster information collection and early warning system is constructed, as shown in Fig. 3.10. The system includes lightning information acquisition devices, current signal and numerical signal conversion devices, storage modules, communication modules, control modules and evaluation and early warning modules. The control module is connected to lightning information collection devices and current signal and numerical signal conversion devices, respectively. The control module is the core to control the functions of other modules. The evaluation and early warning module processes and analyzes the lightning numerical data collected and can monitor the lightning assessment information and early warning information through the communication module. The lightning information acquisition devices are used to monitor the lightning flash frequency and lightning intensity at the location of the embodiment and transmit the monitored information to current signal and numerical signal conversion devices. The current signal and numerical signal conversion device is used to convert the lightning intensity current signal collected by the lightning information device into the numerical signal and transmit it to the storage module. The storage module is used to store the lightning intensity obtained by the lightning information acquisition device and transmit it to the control module for

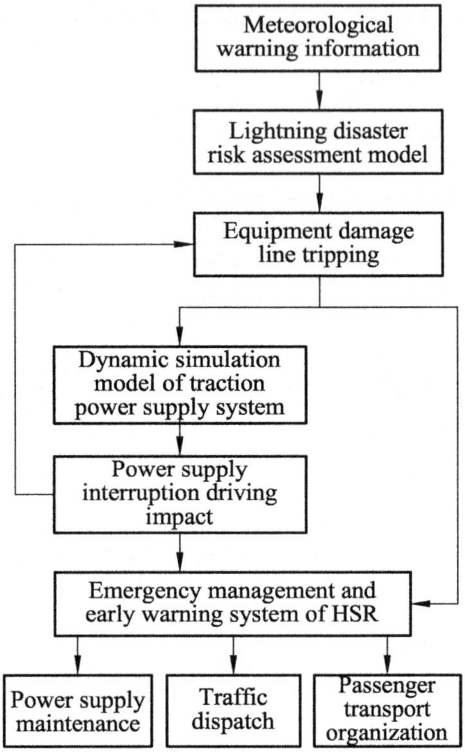

Fig. 3.9 Risk assessment and early warning system of traction power supply system considering the influence of lightning disaster

analysis and processing. The assessment and early warning module is used to dynamically generate the threshold of the lightning impact. When the lightning numerical signal exceeds the threshold, it enters the early warning program, and with the control module, it can send the lightning hazard assessment information and HSR emergency measures through the communication module to realize the safety assessment and early warning process of the lightning disaster for HSR lines. As the external communication module of the information acquisition and early warning device, the communication module transmits the wireless signals to the outside world under the control of the control module or transmits the coded information or short message information through the mobile base station interface to realize communication between the device and the station upper computer terminal, the train cab terminal and passenger mobile phones.

The warning process of the lightning hazard information for the HSR system (see Fig. 3.11) is as follows.

Step 1: Obtain the lightning data. Collect the lightning intensity information at the location of the device and transmit it to the current signal and numerical signal conversion device in the form of a current signal;

Step 2: The SCM converts the incoming current signal into a numerical signal;

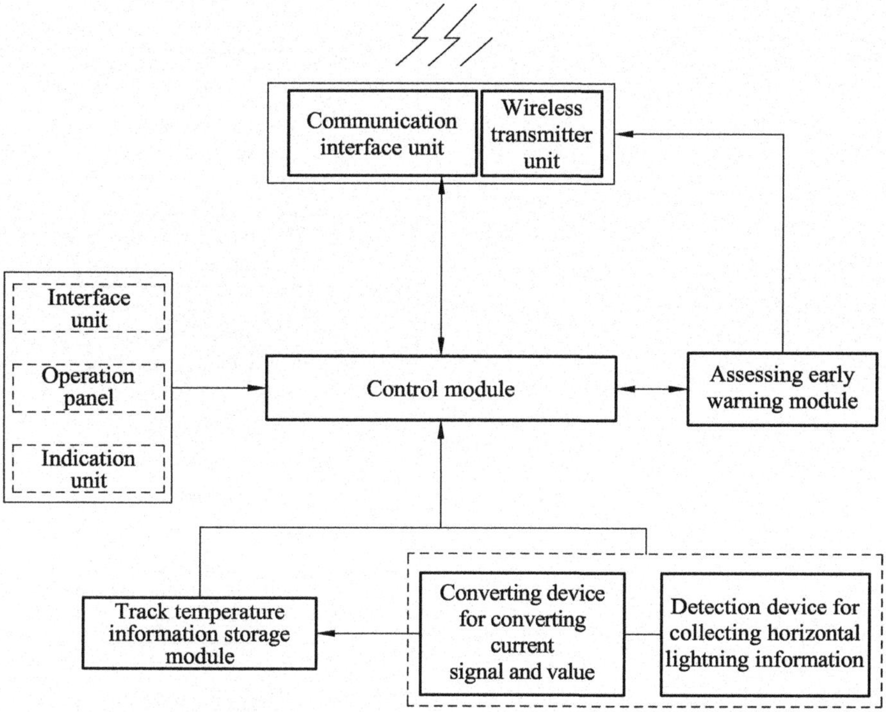

Fig. 3.10 Lightning disaster information collection and early warning system for HSR

Step 3: The control module calls the assessment and warning module to determine the reference threshold interval of the lightning hazard in the HSR line area according to the fixed lightning frequency discrimination threshold;
Step 4: The control module calls the assessment and warning module to determine whether the current lightning numerical signal exceeds the generated lightning hazard threshold;
Step 5: The safety warning device is activated and warning starts;
Step 6: Obtain the spatial coordinates from the lightning information acquisition device and geolocate the affected HSR line;
Step 7: After successful positioning, the management control center where the line is located starts the train control mode;
Step 8: Judge whether the high-speed train is controlled. If the high-speed train is not controlled, return to Step 5 and re-enter the warning. If the train is effectively controlled, enter Step 9 and exit the process;
Step 9: The high-speed train is controlled, exit the process.

The information collection and early warning system for HSR lightning disasters adopts different emergency measures for the control mode of the high-speed

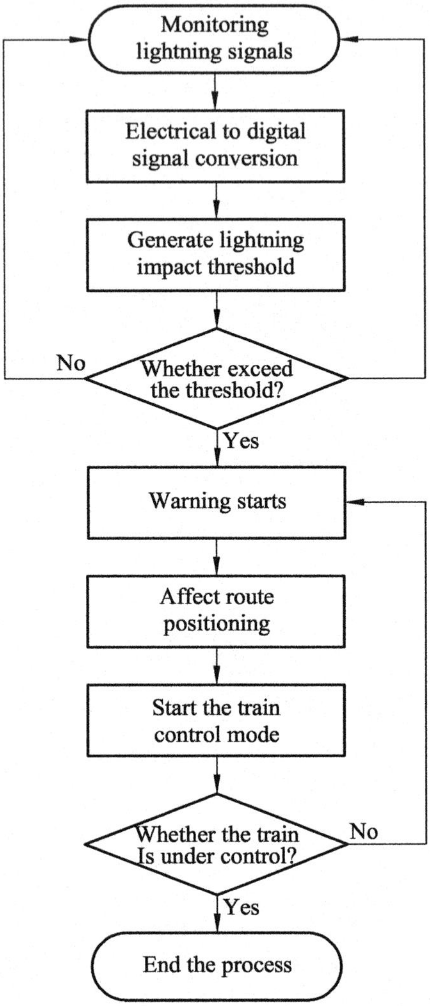

Fig. 3.11 Lightning disaster warning process of HSR

train according to the reference threshold interval of lightning disasters of different levels. Control modes corresponding to different lightning disaster levels are given in Table 3.7.

Table 3.7 Different levels of lightning disaster warning and train control modes

Early warning level		Lightning frequency/[Times/(km^2 min)]	Running speed/(km/h)
Level 1	Blue	15–30	≤ 180
Level 2	Yellow	30–45	≤ 120
Level 3	Orange	45–55	≤ 60
Level 4	Red	> 55	Stop

Table 3.7 shows the different levels of lightning disaster early warning and control modes. The lightning early warning signals for HSR operation safety are divided into four levels, which are respectively expressed with blue, yellow, orange and red.

① **Level 1 of the HSR lightning warning signal: blue warning signal**. The lightning frequency is 15–30 times/km^2 per minute and the running speed is controlled within 180 km/h. Check whether the train power supply and signal equipment operate normally.
② **Level 2 of the HSR lightning warning signal: yellow warning signal**. The lightning frequency is 30–45 times/km^2 per minute, and the running speed is controlled within 120 km/h. Check whether the train power supply and signal equipment operate normally.
③ **Level 3 of the HSR lightning warning signal: orange warning signal**. The lightning frequency is 45–55 times/km^2 per minute and the running speed is controlled within 60 km/h. Check whether the train power supply and signal equipment operate normally, and switch the power supply to the standby battery.
④ **Level 4 of the HSR lightning warning signal: red warning signal**. If the lightning frequency is greater than 55 times/km^2 per minute, the EMU stops.

The information collection and early warning system for HSR lightning disasters has different control modes for different levels of lightning disasters and its early warning system can make the HSR operate more effectively, quickly and safely.

3.4 Summary of This Chapter

In terms of HSR operation safety, lightning strikes are one of the main factors causing the failure of traction power supply systems and signal systems. Due to the sudden and uncertain characteristics of the lightning, the HSR lightning protection system becomes an important part of the HSR protection system. Based on the impact mechanism of lightning, this chapter analyzes the impact mechanism of lightning on HSR and puts forward corresponding evaluation methods. On the basis of summarizing the current situation of the HSR lightning warning in China and abroad, this chapter also analyzes the distribution characteristics and the impact mechanism of lightning disasters and constructs an early warning system for HSR operation safety under lightning disasters in China.

Bibliography

1. Liu L, Yan G, Xin X (2002) Study on schemes and key parameters of seismic alarm system for Beijing-Shanghai express railway. China Safety Sci J 12(4):75–79
2. Wang T, Shi H, He X (2009) Research on emergency response plan for high-speed railway earthquake disaster. Railway Transp Econ 31(8):81–84

3. Arai H, Watanbe I, Motoyama H (2011) Calculation model to evaluate effects of lighting protection measures on railway signaling equipment. In: 2011 international symposium on lightning protection. Fortaleza, Brazil, 247–250
4. Zhao Z, Wu G, Cao X, Li R, Hu J, Zhu J (2011) EGM based calculation method for the installation height of catenary ground wire. China Railway Sci 32(6):89–93
5. Zhai J, Wu W (2021) Travel satisfaction and rail accessibility. Transp Res Part D Transp Environ 100:103052
6. Taniguchi S, Tsuboi T, Okabe S (2010) Improved method of calculating lighting stroke rate to large-sized transmission lines based in electric geometry model. IEEE Trans Dielectr Electr Insul 17(1):53–62
7. Gu S, Feng W, Zhao C, Lu Z (2015) Method of lightning hazard risk evaluation for traction electric network of high-speed railway. High Voltage Eng 41(5):1526–1535
8. Zhou L, Gao F, Li R (2013) Lightning protection system of traction power supply system for high-speed railway. High Voltage Eng 39(2):399–406
9. Glickenstein H (2009) High-speed trains. IEEE Veh Technol Mag 4(4):9–14
10. Wang S, He J, Chen W (2011) Simulation and analysis of lightning overvoltage of 10 kV cable on the bridge of high-speed railroad. High Voltage Eng 37(3):613–622
11. Yu H, Chen S, Zhu J, Cen J (2012) Comparision of lightning location system and artificial thunderstorm days. High Voltage Eng 38(10):2742–2929
12. Wu GN, Gao GQ, Dong AP (2011) Study on the performance of integrated grounding line in high-speed railway. IEEE Trans Power Delivery 26(3):1803–1810
13. Dong AP, Zhang XY, Deng ML (2010) Impact of integrated grounding wire on traction return current in direct power supply system. J Southwest Jiaotong Univ 45(1):88–98
14. Xu Y, He J (2015) Spatial and temporal distribution characteristics of lightning disasters on China's railways. Electr Railway 21(1):47–50
15. Rodrigues RB, Mendes VMF, Catalao JPS (2010) Lighting data observed with lighting location system in Portugal. IEEE Trans Power Delivery 25(2):870–875
16. Zhao L, Yuan M, Tan J, Zhang J, Tang C (2011) Research on lightning shielding characteristics of 500 kV transmission line in plateau mountain. High Voltage Eng 37(6):1663–1669
17. Liu J, Liu M, Qu Z, Xue F, Zhuang Q, Liu H (2010) A new algorithm for the lightning outage rate of the catenary. China Railway Sci 31(2):73–78
18. Ruan L, Gu S, Zhao C, Yao Y, Li X (2012) Technology and strategy of differentiated lightning protection for 220 kV transmission line in three gorges area of western Hubei. High Voltage Eng 38(1):157–167
19. He ZY, Zhang J, Li W (2010) Improved fault-location system for railway distribution system using superimposed signal. IEEE Trans Power Delivery 25(3):1899–1911
20. Li JB, Hu J, Chen YH (2009) Minimum distance of lighting protection between insulator string and line surge arrester in parallel. IEEE Trans Power Delivery 24(2):656–663

Chapter 4
A Temperature Warning System for HSR Operation Safety

High-speed railway adopts a ballastless track structure and interregional seamless line technology, which can facilitate smooth running, low maintenance cost of rolling stock and track structure and long service life of the line. However, the ballastless track and seamless lines also bring a series of technical difficulties. For example, the track structure of the HSR seamless lines will cause expansion due to temperature changes, and the deflection caused by the action of train load is much larger than the general seam track structure, resulting in large expansion amount and flexural force of the seamless line, and the rails are prone to deformation, which affects HSR traffic safety. In the western region of China (Xinjiang, Tibet, etc.), there is not a big temperature difference of rail very year, but the daily temperature difference is big. Repeated changes in daily temperatures may cause brittle failure of track. The northeast region is extremely cold, and the rail will have brittle bending deformation or rail crushing due to the great pressure. Therefore, it is important to monitor and warn the track temperature of the HSR to ensure its operation safety. Based on the research status of the rail temperature early warning system in China and abroad, this chapter studies the impact of high and low temperatures on the track, respectively, and builds an early warning system for HSR operation safety in extreme temperatures.

4.1 Current Situation Analysis for Temperature Disaster Warning System

The ballastless track and seamless line technology not only brings the operating speed of high-speed railway (300 km/h) far beyond that of ordinary railways (150 km/h), but also lead to new technical problems. On the one hand, the longitudinal pressure of the HSR long rails will increase with the rise of rail temperature in hot summer, and the resistance of the track bed will decrease after the operation of the large track maintenance machinery, which will reduce the safety reserve for maintaining the

Q. Hu, *Natural Disaster Warning System for High-Speed Railway Safety Operation*,
Advances in High-speed Rail Technology, https://doi.org/10.1007/978-981-99-7115-2_4

stability of the seamless line. On the other hand, in the extreme cold environment, the cold resistance and frost resistance of the ballastless track roadbed is also a major problem. Therefore, to ensure traffic safety, HSR management departments should control the running speed of the train. The basis of HSR control is mainly on HSR rail temperatures and the lateral resistance of the ballast bed.

4.1.1 Japan's Temperature Warning System

Japan's Shinkansen rail temperature monitoring system consists of rail temperature sensors, atmospheric temperature sensors, stress sensors, signal transmission equipment, information processors, displays, input equipment for the roadbed state information, alarm devices, recorders, information transmission devices and other components, as shown in Fig. 4.1. Rail temperature is closely related to the air temperature, which is almost the same within a small range (generally about 10 km). Therefore, for the ballasted track with curve radius $\leq$ 6 km in Japan, a rail temperature monitoring device should be set every 70 km. In sections with many bridges or curves, such devices should be added appropriately, especially at the beam end of super large continuous girder bridges with large temperature spans. In addition, different speed limits or prohibitions shall be set according to the rail temperature and different ballast bed conditions (such as locked rail temperature, track lifting operation and lateral resistance value) to ensure traffic safety.

The rail locking temperature principle of Japan's Shinkansen seamless line: the difference between the maximum and minimum possible rail temperature and the locking temperature shall not be greater than 40 °C. At the same time, the longitudinal

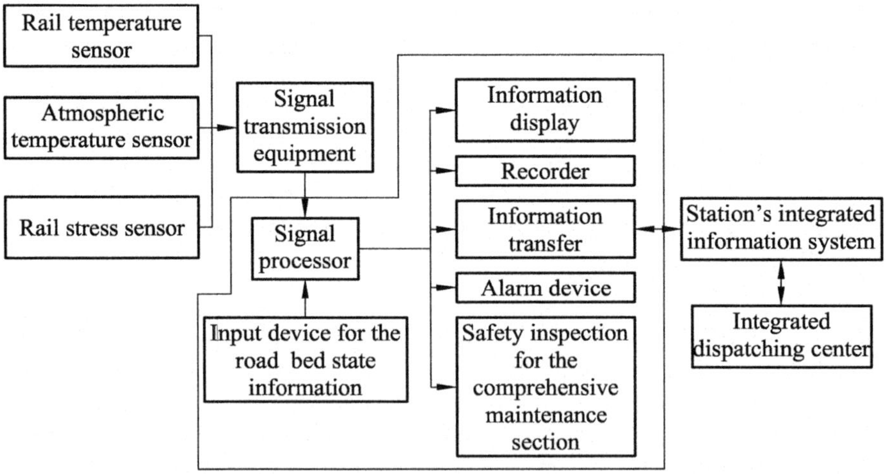

Fig. 4.1 Japan Shinkansen rail temperature warning system

resistance of ballast bed shall not be lower than 8826 N/m. Therefore, Japan defines the maximum rail temperature that can keep the ballast bed stable as about 60 °C and the minimum – 10 °C.

Maximum temperature limit: Considering the track stiffness, the minimum buckling strength to ensure safety rate of 30% is 882.6 kN, and the longitudinal force of the rail when the rail temperature reaches 64 °C is close to 882.6 kN, so 64 °C is determined as the train standstill operation temperature. The longitudinal force with a safety rate of 20% shall be reserved for the minimum buckling strength. The corresponding rail temperature is about 60 °C, and it is required to run slowly at 70 km/h.

Running speed limit: Because the line is constantly subjected to train load, and maintenance operations are repeated, it is possible to change the rail locking temperature or reduce the resistance of ballast bed. To ensure the safety of train operation, it is necessary to monitor the rail temperature and control the train running speed in high temperatures.

Japanese traffic control rules in high temperatures are divided into two categories: the general interval and the ballastless bridge. Each category is managed in accordance with the measured transverse resistance of the ballast bed and the HSR rail temperature. Control standards are given in Tables 4.1 and 4.2.

Table 4.1 Japan Shinkansen driving rules in high temperatures (general interval)

Rail temperature/°C	Measurement of the lateral resistance of the track bed (N/pillow)	Driving rules
> 64	≤ 12,748	Stop running
(60,64)	*< 8826*	Stop running
	≥ 8826	70 km/h slow travel
(58,60)	*< 8826*	Determine whether to travel at 70 km/h according to the special inspection
	≥ 8826	Temperature observation
(53,58)	*< 8826*	Special inspection
	≥ 8826	Temperature observation
(48,53)	*< 8826*	Temperature observation
(45,58)	*< 8826*	A interval special inspection

Table 4.2 Japanese Shinkansen operating rules in high temperatures (telescopic end of ballastless bridge)

Rail temperature/°C	Measurement of the lateral resistance of the track bed (N/pillow)	Driving rules
> 64	≤ 12,748	Stop running
(60,64)	< 10,787	Stop running
	≥ 10,787	70 km/h slow travel
(58,60)	< 10,787	Determine whether to travel at 70 km/h according to the special inspection
	≥ 10,787	Temperature observation
(53,58)	< 10,787	Special inspection
	≥ 10,787	Temperature observation
(45,58)	< 10,787	Temperature observation
(45,58)	< 10,787	B Interval special inspection

Note ① the practice of dispatching personnel to dangerous locations (such as narrow roads and transitional sections of bridges) when the rail temperature reaches a predetermined value. These personnel will enter the first protective fence, patrol outside the protective fence of neighboring lines and visually check the direction and level of the line. ② Section A and B refer to the locations with different longitudinal stability coefficients of the track seam line

4.1.2 Warning System for Temperature Disasters in Britain

Radio-Tech's wireless automatic rail temperature monitoring system has been officially adopted in the British railways. The wireless automatic rail temperature monitoring system consists of the temperature probe installed on the track, the trackside recorder and the monitoring center. With a wireless transmission range of up to 70 m, the temperature probe continuously monitors the rail temperature and transmits measurement data to the recorder at the trackside. The recorder transmits the data to the server of the monitoring center via Internet or Intranet. After processing, the information can be displayed on the screen as a color chart. When the temperature reaches the critical point, the alarm shall be promptly raised. The monitoring center shall dispatch the personnel to monitor closely and take the necessary measures in time. Both the temperature probe and the recorder are battery-powered. The recorder battery has a lifespan of 1 year, and the temperature probe battery can be used for 10 years. The wireless automatic rail temperature monitoring system does not require wiring and can be employed at bridges, tunnel entrances and wooded areas. The system components are shown in Fig. 4.2.

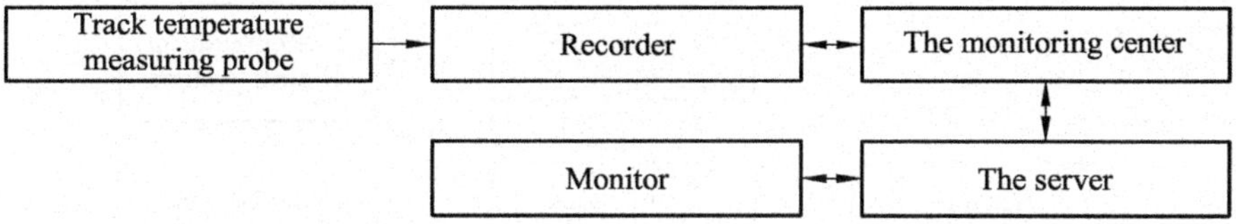

Fig. 4.2 Wireless automatic track temperature monitoring system

4.1.3 Warning System for Temperature Disasters in Germany

German high-speed railways are exposed to direct sunlight, with a large daily temperature differences in summer (up to 40 °C) and a modest temperature differences in winter (only about 12 °C), making an HSR rail temperature detection system essential. German HSR adopts a new disaster prevention alarm system—MAS90, which can not only track the operational status of line equipment, but also spot and promptly notify the impact of the environment on the traffic safety, as well as the damage to the mobile equipment. The new disaster prevention alarm system is equipped with a central control unit (SZE) in the south, north and middle sections of the whole line, which are interconnected; each SZE also connects various alarm and recording units (MRE) located in the signal building along the line and communicates with them to exchange data and instructions. MRE receives the information collected by the detection and alarm instruments installed along the line. The German rail temperature monitoring subsystem is shown in Fig. 4.3.

The rail temperature difference is also regulated in Germany. The German *Code for Railway Bridges and Other Engineering Structures (DS804)* stipulates that when calculate the longitudinal forces in special cases, the variation of the rail temperature should be considered in accordance with the corresponding guidelines in the *Superstructure Regulations Manual (DS802)* for standard gauge railways. In order to simplify the calculation, a symmetrical temperature variation of ± 50 °C is allowed.

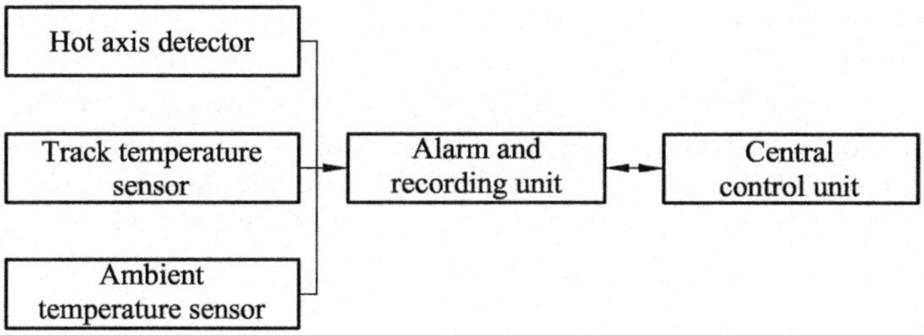

Fig. 4.3 Early warning system for safe operation of HSR in temperature environment in Germany

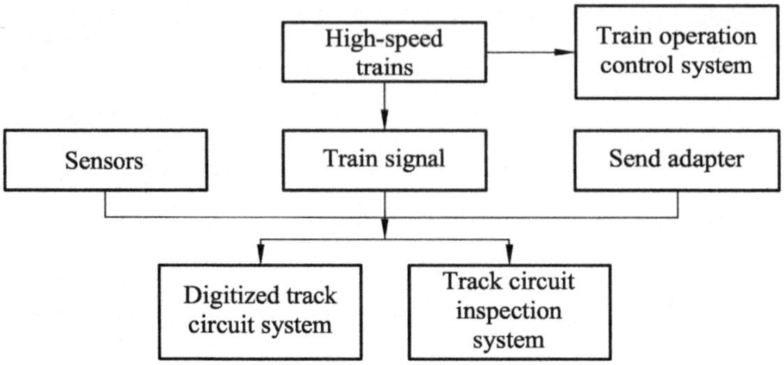

Fig. 4.4 Automatic train control system of TGV

4.1.4 Warning System for Temperature Disasters in France

French HSR adopts the automatic train control system which is mainly based on locomotive signals. In the model TVM430 which is an automatic train control system (ATC), in addition to the automatic control of train speed, equipment status and natural environment monitoring and alarm subsystems are added, it further strengthens the guarantee function of train operating safety. Seven different subsystems are part of its automatic train control, including automatic train detection, catenary voltage detection, hot box detection, rainfall detection, snowfall detection, high wind detection, and overpass falling item detection, among others. Figure 4.4 shows the screenshot of TVM430 automatic train control system.

The French HSR rail temperature monitoring system mainly installs rail temperature monitors on the ground along the HSR line in order to monitor the axle box temperature and changes of all passing vehicles. An early warning signal will be sent right away, but the speed will not be limited if it is discovered that the temperature of one or more axle boxes exceeds the initial temperature threshold value. When the second temperature threshold is exceeded, the alarm shall be sent to the dispatching command center through the radio transmission system or TVM430 system, the train shall be required to stop in an emergency, while notifying neighboring lines on the train speed limit, and the neighboring lines on the train speed limit shall be notified at the same time. To guarantee the high reliability of the rail temperature monitoring system for French HSR, one ground axle temperature monitor is placed at a certain distance.

4.1.5 Temperature Warning System of China's Harbin HSR

China's Harbin–Dalian (Harbin to Dalian) high-speed railway line, which started construction in July 2007 and opened for operation in December 2012, is the world's first HSR to operate in a severe cold area. To ensure the safe and smooth operation,

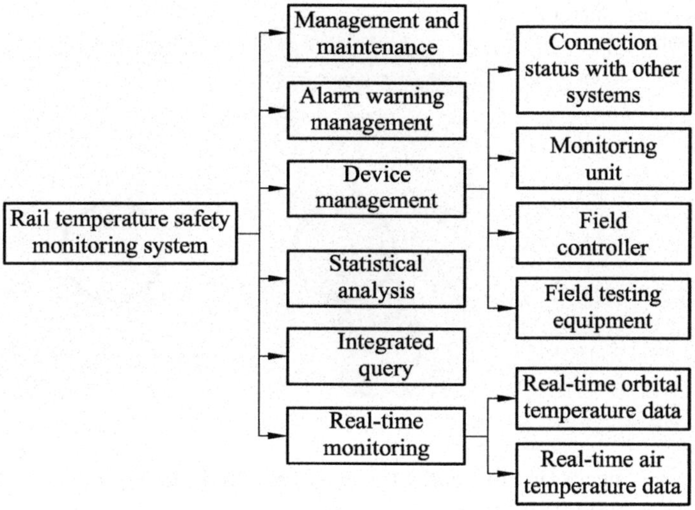

Fig. 4.5 China's HSR temperature warning system

the rail temperature needs to be monitored constantly and warned in time. China's HSR temperature monitoring system consists of the safety management system, the disaster prevention and safety monitoring system of passenger dedicated lines. It also exchanges and shares information with dispatching command, emergency rescue, traffic safety monitoring, passenger service, comprehensive maintenance, traction power supply and train control system. The overall structure of the HSR temperature monitoring system is shown in Fig. 4.5.

4.2 Influence Mechanisms of the Temperature on the HSR Safety Operation

The HSR track system consists of steel rails, fasteners, rail sleepers, road beds and other components. The HSR track system is the basis of operating. Steel rail is the main component of the HSR track, whose role is to guide the high-speed train wheels forward, bear the huge pressure of the wheels and transfer to the rail sleeper. In the past, the rail with standard length of 25 m was used for railway lines. Adjacent rails were connected by joint cleats, and a few millimeters to more than ten millimeters of rail gap were left. However, HSR lines generally adopt long rails without gaps. Seamless rail lines can ensure the smoothness of the train on the way and reduce the damage to the track components. As the rise and fall of the temperature will lead to the expansion and contraction of the rail, which will be converted into temperature stress, resulting in the deformation of the track. In order to ensure the HSR safety

Fig. 4.6 Seamless rail for high-speed railway

operation, it is necessary to conduct an in-depth study of the mechanism of how temperature would affect the track.

4.2.1 Characteristic Analysis of High-Temperature Disasters

As the high-speed railway is paved with cross-sector seamless line, the longitudinal stress of the long rails of the seamless line will increase with the rise of the rail temperature in summer. If the large-scale road maintenance machinery is operated at night in this season, the state of the ballast bed before the operation will be changed after the operation. The actual measurement shows that the longitudinal and transverse resistance of the ballast bed has decreased, and the safety reserve for the stability of the seamless line will be reduced at this time. If the rail temperature continues to rise to (or beyond) a certain critical value, as long as there is any disturbance (such as vibration when passing the train, braking in the section and line maintenance), the seamless line will lose its ability to maintain stability and thus cause buckling accidents (Fig. 4.6), posing a threat to the HSR safety operation.

The HSR rail temperature forecasting and monitoring system can monitor the rail temperature, safety reserve, weather and other information of seamless lines in real time, providing decision basis for the HSR engineering maintenance department and comprehensive dispatching center.

4.2.2 Characteristic Analysis of Low-Temperature Disasters

The operation conditions of high-speed railway have high requirements on the subgrade, and even minor changes will bring disastrous consequences. Therefore, the particularity of seasonal frost heaving in the ballastless track subgrade structure

Fig. 4.7 High-speed railway subgrade

is one of the key technical problems that perplex the construction and operation of HSR in severe cold areas. The HSR in the cold areas has two technical problems: one is the force characteristics of the HSR under the rail, which needs to solve the strength and stability of the foundation under the rail; the other one is the deformation characteristics of subgrade during seasonal freezing and thawing, and the problem of subgrade frost heaving needs to be solved (Fig. 4.7).

Subgrade frost heaving refers to the phenomenon that under the conditions of negative temperature and certain temperature gradient, the water in the subgrade body migrates to the frozen surface and freezes and the subgrade soil volume increases after freezing, resulting in the change of the elevation of the subgrade top. The environment temperature is one of the important factors affecting soil frost heaving. The freezing and thawing of soil are mainly affected by the environment temperature. Therefore, it is very important for HSR to establish the seamless rail temperature forecasting and monitoring system and transmit the data to the safety and disaster prevention alarm system.

4.2.3 Definition of Parameters for the Temperature Monitoring and Warning

In order to study the impact of the HSR rail temperature, the temperature stress and rail temperature of HSR rails are taken as the main indicators for quantitative research to determine the threshold value of the temperature impact on the normal HSR operation. Based on it, the safety interval of HSR speed under different temperatures is determined, which provides a theoretical basis for the design of the HSR rail temperature monitoring and forecasting system.

(1) **Rail temperature**. The HSR rail temperature is referred to as rail temperature for short, which is different from the air temperature. Many factors affecting the rail temperature are related to climate change, wind power, sunlight intensity, rail location and measurement location. HSR construction is now generally using

ultra-long seamless rails, whose advantage is to reduce the impact of vibration during the operation of the train. HSR rail temperature stress is caused by the rail temperature. In the initial stage of seamless rail erection, the internal stress is zero. After the laying of the rail sleeper, constraints are formed on the HSR rails. The extra-long seamless rails cannot be deformed freely and the rise or fall of the HSR rail temperature will cause temperature stress on it.

In order to ensure the traffic safety, the rail temperature must be limited to a certain range, so as to avoid the deformation of HSR rails. The high temperature in summer is easy to cause rail expansion, while the cold temperature in winter makes it impossible to contract normally, which is easy to cause rail breakage.

Range of the HSR rail temperature: The maximum rail temperature is generally higher than the maximum local temperature of about 20 °C, while the minimum rail temperature and the minimum local temperature is roughly the same.

(2) **Warning threshold**. The size and distribution of the rail temperature stress of the HSR seamless line has a direct relationship with the range of the rail temperature change, which is the main factor affecting the strength and stability of the seamless line. Therefore, the range of the rail temperature has become an important information for the design, laying, maintenance and repair of the seamless line. High-speed railroad rails use 60 kg/m and 100 m fixed-length rails.

If the temperature variation of the HSR rail is Δt, the expansion of a rail of length l is

$$\Delta l = \alpha l \Delta t. \tag{4.1}$$

The strain generated by the HSR rail is

$$\varepsilon = \Delta l / l = \alpha \Delta t. \tag{4.2}$$

If the HSR rail cannot be retracted due to resistance, the generated Δl will be converted into stress, according to Hooke's law:

$$\sigma = E\varepsilon. \tag{4.3}$$

Thereby the temperature force on the HSR rail is

$$P = \sigma S, \tag{4.4}$$

where

α is the linear expansion coefficient of the HSR rail;
E is the elastic modulus of the HSR rail;
S is the cross-sectional area of the HSR rail.

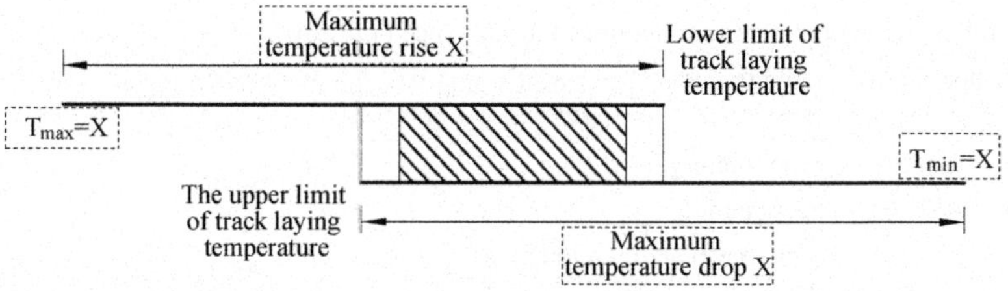

Fig. 4.8 Setting diagram of the locked rail temperature

Warning interval of the HSR rail temperature: In the locked HSR rail, with the increase of temperature load, the longitudinal temperature force inside the HSR rail internal is also increasing and is in direct proportion to the temperature variation. The maximum allowable temperature rise, maximum allowable temperature fall, extreme maximum and minimum rail temperature in local history shall be considered in the calculation of locked HSR rail temperature.

$$[\Delta t_d] + [\Delta t_c] = 149.05 > T_{\max} + T_{\min} + 10 = 96\,^{\circ}\mathrm{C}, \quad (4.5)$$

where

$[\Delta t_d]$ is the maximum allowable temperature fall;
$[\Delta t_c]$ is the maximum allowable temperature rise;
$T_{\max}$ is the extremely high rail temperature in local history;
$T_{\min}$ is the extremely low rail temperature in local history.

Therefore, the locked rail temperature of the temperature strained CWR track shall ensure that the runway will not expand in summer and the rail will not break in winter. The locked rail temperature is set according to Fig. 4.8.

Warning criteria of the HSR rail temperature: Based on the operating regulations of the Japanese Shinkansen in high-temperature conditions, (Table 4.3) and (Table 4.4) respectively provide the operating regulations for China's high-speed railway under high and low temperature conditions.

4.3 Temperature Warning System for the HSR Safety Operation

The structure of HSR rail temperature real-time monitoring and early warning device is shown in Fig. 4.9. The HSR rail temperature real-time monitoring and early warning system is made up of temperature information acquisition module, current signal and numerical signal conversion device, storage module, communication module, control module, evaluation and early warning module and other components. The control module is respectively connected with the temperature

Table 4.3 High-temperature running rules for high-speed railway

Rail temperature/°C		Driving rules
Level 1	> 64	Stop running
Level 2	(60,64]	Stop running
		Slow travel at 70 km/h
Level 3	(55,60]	Determine whether to travel at 70 km/h according to the special inspection
		Temperature observation
Level 4	(51,55]	Special inspection
		Temperature observation
Level 5	(45,51]	Temperature observation
Level 6	(42,45]	Special inspection of interval A

Table 4.4 Low-temperature running rules for high-speed railway

Rail temperature/°C		Driving rules
Level 6	[− 55,− 50)	Stop running
Level 5	[− 50,− 45)	Stop running
		Slow travel at 70 km/h
Level 4	[− 45,− 40)	Determine whether to travel at 70 km/h according to the special inspection
		Temperature observation
Level 3	[− 40,− 35)	Special inspection
		Temperature observation
Level 2	[− 35,− 30)	Temperature observation
Level 1	≤ − 30	Special inspection of interval A

information acquisition module, current signal and numerical signal conversion device, storage module, communication module, control module, evaluation and early warning module. As the core module, the control module is used to realize the functions of the above modules. The control module and the evaluation and early warning module process and analyze the collected temperature information numerical data. It can also transmit the assessment information of the rail temperature disaster through the communication module. The early warning process of the HSR rail temperature information collection and monitoring method is shown in Fig. 4.9.

As can be seen from Fig. 4.9, the temperature information acquisition module of the HSR rail temperature real-time monitoring and early warning system is used to monitor the rail temperature and environment temperature at the location of the device and extract the spatial coordinate information. It can also record the monitoring information and spatial coordinate information in the storage module.

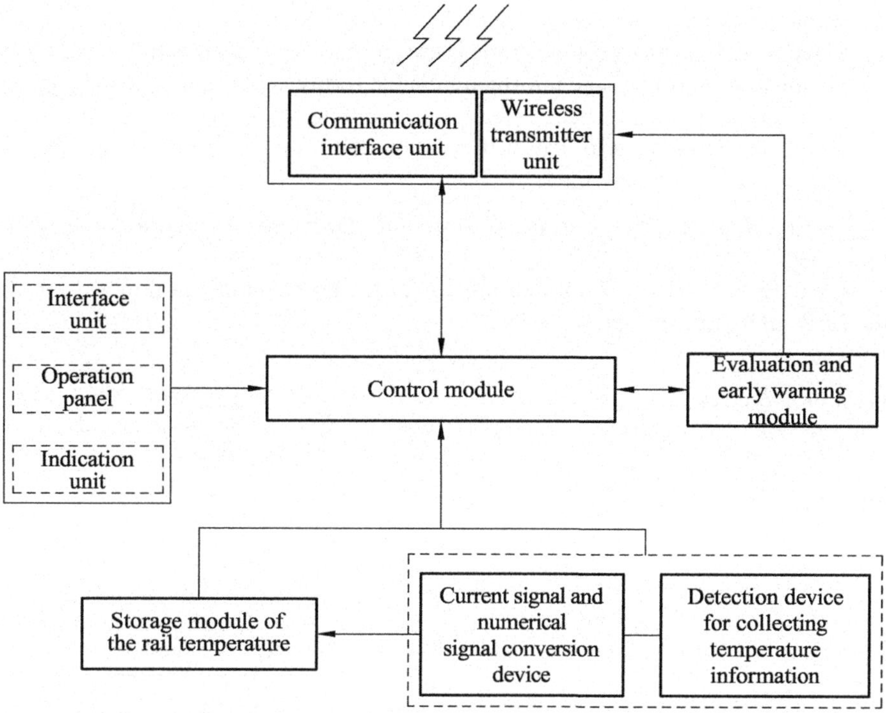

Fig. 4.9 Structure of temperature warning system

① The current signal and numerical signal conversion device of the temperature warning system. It is used to convert the current signals of rail temperature and environment temperature collected by the temperature information device into numerical signals, which will be transmitted into the storage module.

② The storage module of the temperature warning system. It is used to store the aforementioned rail temperature parameter information and spatial location information, and can transmit the rail temperature parameter information through the communication module under the control of the control module.

③ The evaluation and early warning module of the temperature warning system. It is used to evaluate the collected rail temperature parameter information and determine the safety level of the current rail temperature according to the dynamically generated hazard threshold of the rail temperature. The module will enter the early warning procedure when judging the rail temperature value signal exceeds the threshold and can send the disaster evaluation information of rail temperature and the HSR emergency measures through the communication module under the control of the control module to realize the safety assessment and early warning process of the HSR rail temperature.

④ The communication module of the temperature warning system. As the external communication module of the temperature information acquisition and early warning device, under the control of the control module, it sends wireless signals to the outside world through wireless transmission or transmits the coded information and short message information in the form of short message through the mobile base station interface to realize the communication among the device and the upper computer terminal of the station, the train cab terminal and the cell phone terminal of passengers.

The flowchart of the early warning system for HSR rail temperature is shown in Fig. 4.10.

Based on Fig. 4.10, the warning process of the temperature warning system for the HSR safety operation is as follows.

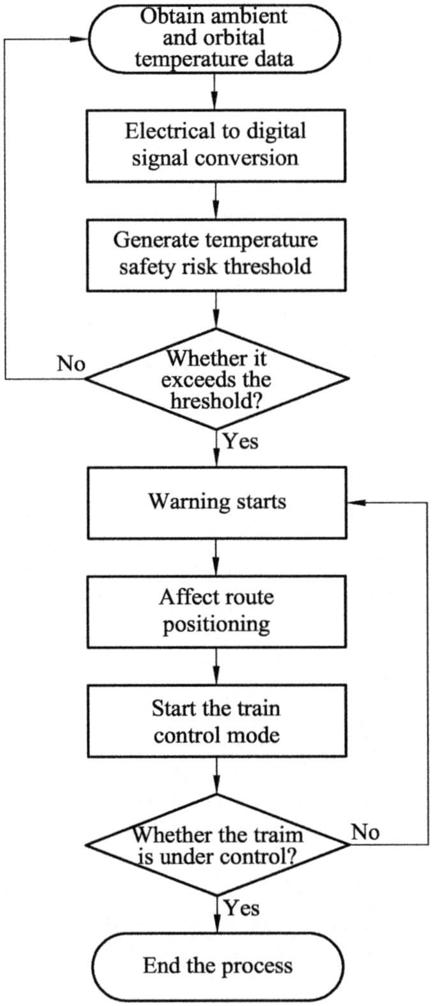

Fig. 4.10 Process of the HSR temperature early warning system

Step 1: Obtain the HSR rail temperature parameter information, and incorporate the current signal into the current signal and numerical signal conversion device;
Step 2: The SCM converts the incoming current signal into a numerical signal;
Step 3: The control module calls the evaluation and early warning module to compare the historical data of the rail temperature disaster with the actual situation of the current area where the example is located, and determine the reference threshold range of rail temperature disaster in the high-speed railway line area;
Step 4: The control module calls the evaluation module to analyze the preprocessed rail temperature and environment temperature parameter information to determine whether the current rail temperature fluctuation numerical signal in the area where the implementation example is located exceeds the generated rail temperature disaster threshold value;
Step 5: Start the safety warning device and start the warning;
Step 6: Obtain the spatial coordinate information where the rail temperature information acquisition device is located, and locate the affected HSR lines;
Step 7: After successful positioning, the management control center where the HSR line is located starts the train control mode;
Step 8: Judge whether the train is under control. If the train is not under control, return to Step 5 and re-enter the warning system; if the train is effectively controlled, go Step 9 and exit the whole process;
Step 9: When the train is under control, exit the process.

4.4 HSR Rail Temperature Warning System

The HSR rail temperature is crucial to the stability of the seamless line, and the real-time monitor of the rail temperature of the seamless line can effectively prevent the occurrence of rail expansion and rail breakage. Especially in the northeast and western areas of China (Xinjiang, Tibet, etc.), there is a large area of permafrost region, which seriously affects the safety of the HSR. It is important to establish a real-time rail temperature monitoring system along the HSR line and conduct long-term, comprehensive and real-time observation of the HSR track to ensure the safe operation of the HSR line.

4.4.1 Early Warning Content of the HSR Rail Temperature

The rail system is an essential part of the HSR system. HSR lines in China generally use long rails without any gaps. The expansion of the rail caused by the rise and fall in the seamless rail temperature will convert into temperature stress, which will lead to deformation of the track. The traditional method of the rail temperature monitoring is to use human labor to measure the rail temperature at a fixed time and location. If the temperature of air and rail go beyond the limit, it is necessary to patrol and monitor the line changes. This method is time-consuming, laborious, and inefficient. It is unable to realize real-time monitoring and real-time warning. It is also unable to take the necessary actions for the section whose rail temperature breaches the threshold for the first time. If the HSR rail temperature reaches (or exceeds) a certain critical value, while there is still excitation, the seamless line will be unable to maintain stability, leading to rail expansion and runway incidents and endangering the safety of the HSR. Therefore, installing the rail temperature monitoring and forecasting system along the HSR is an important means to ensure the HSR safety operation. At present, the research on the rail temperature monitoring along the HSR in Japan is currently in a relatively advanced stage. Based on Japanese study findings, the data content of HSR rail temperature monitoring in this chapter include:

① HSR rail temperature. The HSR construction now generally uses extra-long seamless rail, which is easily responsible for rail expansion under high temperatures in summer; in the severe cold weather in winter, it is impossible for it to shrink normally, and rail breaking is easily caused.

② HSR ambient temperature. The ambient temperature is the temperature of the HSR operation environment, generally referring to the external temperature. According to a significant number of long-term measurements, the maximum rail temperature is generally around 20 °C higher than the local maximum temperature, and the minimum rail temperature is roughly equal to the local minimum temperature.

4.4.2 Data-Aware Model for the Rail Temperature Warning

Support vector machine (SVM) is a novel machine learning method proposed by Vapnik et al. in the early 1990s, based on statistical theory. SVM is based on the theoretical principle of structural risk minimization. By appropriately selecting the function subset and the discriminant function in that subset, the actual risk of the learning machine is reduced, guaranteeing that the small error classifier obtained by a limited training sample still has a small test error for the independent test set. This chapter analyzes the HSR rail temperature monitoring data based on SVM.

(1) **The basic principle of SVM**. The main principle of the SVM is to find the optimal classification hyperplane of two types of samples in the original space under the linearly separable condition; in the case that the two types of samples are linearly indivisible, the relaxation variable is introduced for analysis. By using nonlinear mapping, the samples in the low dimensional input space are mapped to the high-dimensional attribute space to make it linear, so that it is possible to analyze the nonlinearity of samples in the high-dimensional attribute space by using linear algorithms, and find the ideal classification hyperplane in the feature space. By utilizing the structural risk reduction concept, SVM creates the ideal classification hyperplane in the attribute space, ensuring that the classifier is globally optimal, and the predicted risk in the entire sample space meets a specific upper bound with a certain probability.

The traditional way of detecting temperature in China's railway system in the past was carried out manually. In general, the testing frequency was higher during the warm season and lower during the winter. In the environment of northeast and western regions of China, the manual measurement of rail temperature is a difficult work that is susceptible to inclement weather, wastes of human and material resources and with low reliability. At present, this measurement is unable to meet the need of large-scale observation along the line, so it is necessary to monitor the HSR along the line for its entire lifespan and carry out the monitoring and forecasting of the rail temperature to ensure the HSR safety operation. Therefore, this chapter applies the fundamental SVM idea to build a HSR rail temperature monitoring and forecasting system in China.

(2) **The basic algorithm of SVM**. The SVM is essentially a 2-value classifier, and the classification decision function is as follows:

$$\begin{aligned} y &= f(x) \\ &= \operatorname{sign}\left(\sum_{i=1}^{n} \alpha_i y_i K(x_i, x) + b\right) \\ &= \operatorname{sign}\left(\sum_{\forall \alpha_i \in \mathrm{SV}} \alpha_i y_i K(x_i, x) + b\right), \end{aligned} \tag{4.6}$$

where

$K(x_i, x)$	is the kernel function;
x	is the sample to be classified;
(x_i, y_i)	is the training sample set, and $i = 1,2, \ldots, n$;
n	is the number of training samples;
$x_i \in R^d$	is the training sample;
$y_i \in \{+1, -1\}$	is the class label of the sample;
SV	is the support vector set which is a subset of the training sample.

Therefore, in order to obtain the maximum information based on the limited data, the HSR rail temperature monitoring and forecasting system can be created by employing SVM. Different support vector algorithms are proposed, and different kernel functions can be utilized to build learning machines for various kinds of nonlinear decision surfaces in the input space. A basic function relation is adopted in the early warning system of the HSR rail temperature.

$$\begin{aligned} &\max : L_D = \sum_{i=1}^{n} \alpha_i - \frac{1}{2} \sum_{i,j=1}^{n} \alpha_i \alpha_j y_i y_j K(x_i, x_j), \\ &\text{s.t.} \begin{cases} C \geq \alpha_i \geq 0 \\ \sum_{i=1}^{n} \alpha_i y_i = 0 \end{cases}, \end{aligned} \tag{4.7}$$

where C is a predetermined constant that controls the degree of penalty for misclassified samples, striking a balance between the proportion of misclassified samples and the complexity of the algorithm. The penalty power $C > 0$ regulates interclass overlap, whereas $C = 0$ forbids it.

The current kernel functions commonly used in support vector machines are as follows:

Linear kernel function $K(x, x_i) = (x, x_i)$;
Polynomial kernel function $K(x, x_i) = [(x, x_i) + 1]^q$;
Gaussian kernel function $K(x, x_i) = \exp\{-(x - x_i)^2/\sigma^2\}$;
Sigmoid kernel function $K(x, x_i) = \tan(v(x, x_i) + c)$.

In this chapter, the HSR rail temperature warning system is implemented using the linear kernel function to simplify the process.

4.4.3 Early Warning System for the Rail Temperature

In winter, the conductive rail and ballast of the HSR could freeze; in summer, the HSR track could expand due to extreme high temperatures. If measures are not taken in time, the train may derail. Currently, the monitoring of HSR sections is accomplished by setting up observation stations along the HSR tracks, installing temperature sensors and data collection units. Additionally, when the temperature of the HSR exceeds the alarm threshold, the temperature data is promptly collected and an alarm is triggered. Users should take timely response measures, such as slowing down, stopping, or taking evasive actions, after verifying the alarm information and assessing the site conditions. The users of disaster prevention systems are generally railway dispatchers. If the high temperature, especially the extremely high temperature, are warned in advance before the high temperature affects the daily operations and the decision-making time is allowed for dispatchers to determine reasonable and appropriate countermeasures, it will play a very good role in preventing disasters and ensuring the safety and efficiency of traffic.

(1) **Composition of the early warning system**. Through the comparison of domestic and international rail temperature monitoring and early warning system, the HSR rail temperature monitoring and prediction system is designed. The early warning system includes:

Component I. By using temperature sensors, the design of environmental temperature and rail temperature information collecting interface is implemented;

Component II. Through 802.11-WIFI wireless transmitter module, the real-time transmission of the sensor data is realized;

Component III. Based on Visual Studio platform, the upper computer data fusion processing center is designed;

Component IV. Based on the traditional warning and discernment of the rail temperature, taking into account the impact of the HSR rail temperature on train operation, the assessment algorithm of the HSR safety operation based on the rail temperature is designed;

Component V. Based on the mobile communication base station, the pushing function of the assessment information for the HSR safety operation is realized.

The HSR early warning system based on the rail temperature detection can be specifically divided into three parts: threshold discrimination, system structure, integration testing, etc.

(2) **Judgment of the Early warning threshold**. According to the monitoring interval of the HSR rail temperature: the maximum rail temperature is generally about 20 °C higher than the local maximum temperature, and the minimum rail temperature is roughly equal to the local minimum temperature. With reference to existing research findings at home and abroad, the HSR driving rules under high temperatures are determined as given in Table 4.5.

The discriminating threshold of rail temperature along the HSR is determined in accordance with the HSR traffic, as given in Table 4.6.

(3) **Structure of the early warning system**. The structure of the device for real-time monitoring and early warning of the HSR rail temperature is shown in

Table 4.5 Rules for China high-speed railway train in high temperatures

Rail temperature/°C		Driving rules
Temperature interval	Warning threshold	
> 65	65	Stop running
(60,65]	60	Stop running
		Slow travel at 70 km/h
(55,60]	55	Determine whether to travel at 70 km/h according to the special inspection
		Temperature observation
(50,55]	50	Special inspection
		Temperature observation
(45,50]	45	Temperature observation
(40,45]	40	Special inspection of interval A

Table 4.6 Discriminating threshold of track temperature along high-speed railway

Track temperature along HSR(°C)	Speed limit for train operation
[0,45]	Normal operation
(45,51]	The speed limit of 300 km/h
(51,55]	The speed limit of 200 km/h
(55,60]	The speed limit of 120 km/h
> 60	Stop running

Fig. 4.11, including temperature information acquisition, current signal and numerical signal conversion device, storage module, communication module, control module, evaluation and early warning module. The control module of the HSR rail temperature is connected to the temperature information acquisition module, current signal and numerical signal conversion device, storage module, communication module, control module, and evaluation and warning module, respectively. The control module of the HSR rail temperature is utilized as the core module to control the function of the above modules. The control module and the evaluation and warning module process and analyze the numerical data of the collected temperature information and can convey the evaluation information of the rail temperature disaster through the communication module.

(4) **Test of the early warning system**. In order to verify the stability and practicality of the system, virtual HSR rail temperature is used to test the HSR early warning system based on the rail temperature detection. The test method procedure involves first gently heating the track, followed by testing the vibration of the HSR rail temperature information, environmental temperature information and warning information collected by the entire system throughout the vibration process of the HSR rail temperature. In the actual test, the warning level is divided into three levels. The final test results obtained are shown in Fig. 4.12.

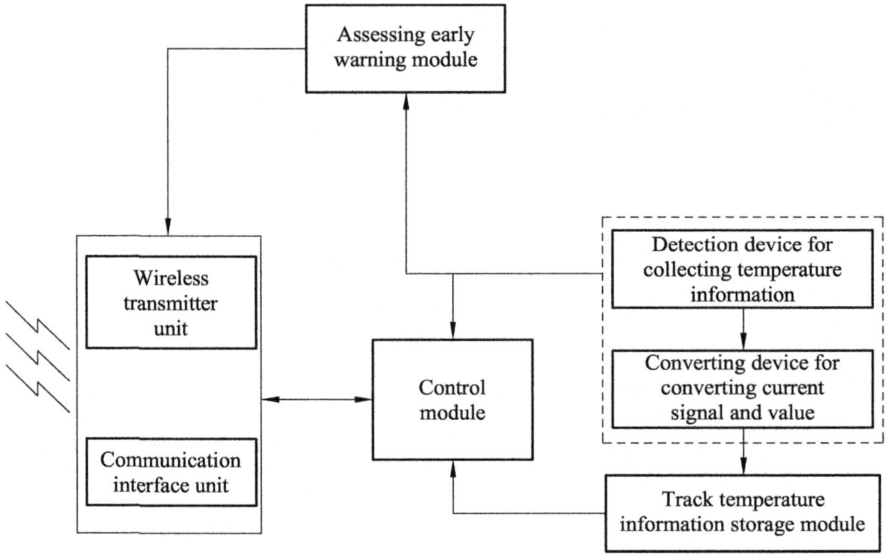

Fig. 4.11 Structure of temperature monitoring system

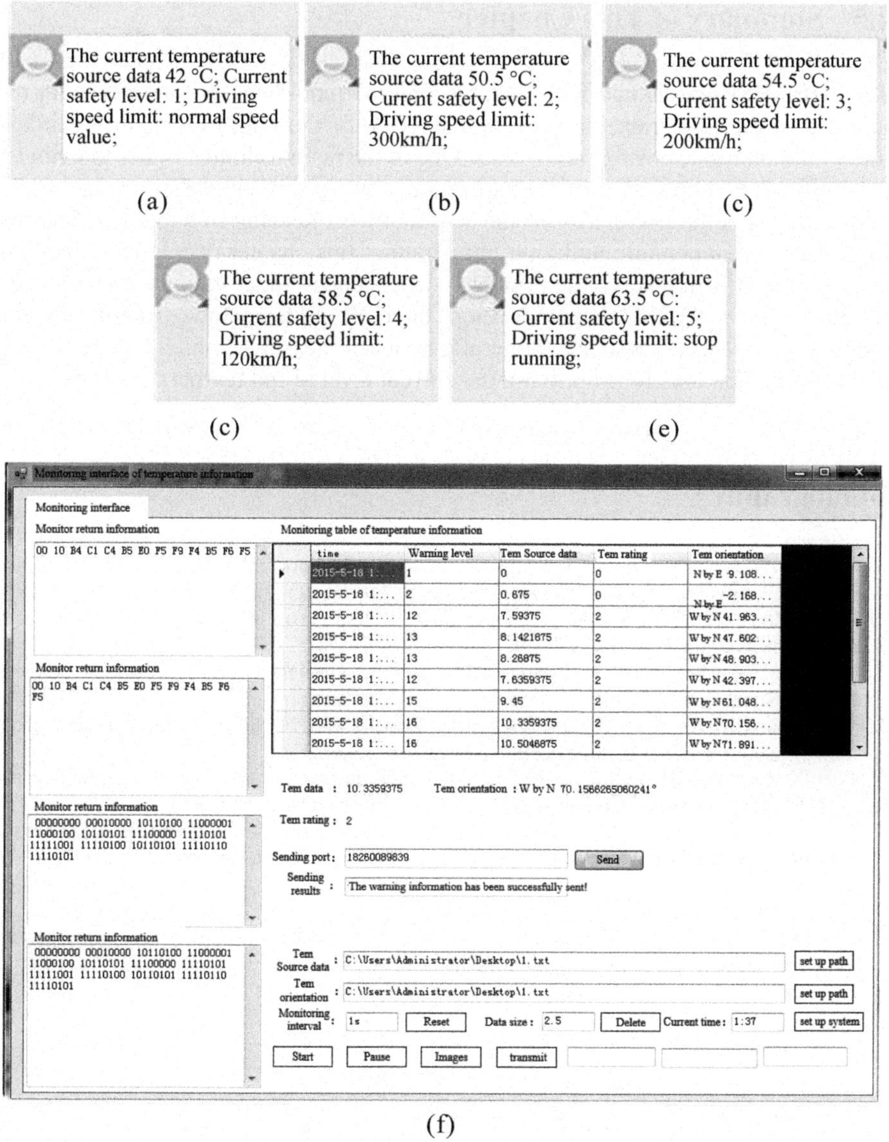

Fig. 4.12 Test results of HSR early warning system based on the rail temperature detection. *Note* The above picture is the original Chinese interface

4.5 Summary of This Chapter

The entire HSR line is laid with seamless interregional lines. In summer, with the increase of the HSR rail, the longitudinal stress of the long rails of the seamless line would increase, posing safety risks. One of the main technical issues that hinder the construct and operation of high-speed railways in some high latitude and high-altitude areas is the peculiarity of the seasonal frost heaving of subgrade. Starting from the low temperature and high temperature, this chapter deeply examines the stress of the HSR track and the possible consequences under various temperatures. Based on this theory, this chapter develops the HSR rail temperature monitoring and prediction system and associated operation control measures, which has significant practical implications for improving the current level of rail temperature detection.

Bibliography

1. Chen X, Ye L (2009) Design and implementation of network video surveillance system for ningbo rail transit project construction. Railway Stand Des 22(12):130–133
2. Huo H, Zhang H, Zhang S (2007) Research on a wireless sensor network used in railway track inspection. J Xidian Univ 34(1):35–38
3. Wang Y, Wang S (2015) Monitoring technique for ballestless track of high-speed railway. Railway Stand Des 59(8):1–9
4. Wu S, Wang X, Wu Z, Lu F (2013) Structural damage types of ballastless track in high-speed railway and rapid repair method. Chin Railways 51(1):42–44
5. Chen X, Zhao W (2013) Influence of bridge temperature span on stress state and on reinforcing bars of longitudinally-continuous base layer. Railway Stand Des 15(10):6–9
6. Chakmbarti A, Sabharwal A, Aazhang A (2006) Communication power optimization in a sensor network with a path-constrained mobile observer. ACM Trans Sensor Netw 2(3):297–324
7. Ma Y (2007) Application of high-tech in railway video surveillance field. Railway Sign Commun Eng 4(1):26–29
8. Qi F (2014) Application and practice of high-speed rail disaster prevention system in intercity railways. Shanghai Railway Sci Technol 18(1):115–117

Chapter 5
A Rainstorm Warning System for the HSR Safety Operation

Railway transportation is less affected by weather and other environmental factors than other forms of transportation and can ensure the continuity and punctuality of operation. Both the safety and punctuality of high-speed train operation are seriously affected by the natural environment, especially the impact of rainstorm on the HSR safety operation. The impact of heavy rainfall on the HSR operation is mainly manifested in: On the one hand, the railway department usually takes control measures to slow down or stop in a light rainfall, which causes the delay of the HSR and affects the travel efficiency of passengers; On the other hand, debris flow caused by heavy rainfall has a more serious impact on the HSR operation (see Fig. 5.1). Therefore, the effective use of the disaster prevention and warning system to improve the emergency response level of traffic in the severe weather is an important assurance to lower the safety risk of flood and disaster, reduce the number of HSR delays and decrease the impact scope of delays.

The HSR rainfall monitoring and early warning system works in real time to prevent impacts of landslides, debris flows, dangerous rocks, falling rocks and collapses on the HSR traffic safety in risky mountain areas and subgrade sections along the railway during the flood season. The HSR rainfall monitoring system not only has the function of analyzing and processing the rainfall data from both intermittent and continuous rainfalls, but also has the function of setting rainfall alarm release authority on the monitoring terminal, specifically by adjusting the alarm level, alarm threshold, control range and other parameters. Therefore, the HSR rainfall monitoring and early warning system primarily assists the departments like track maintenance division and flood prevention office of each railroad bureau and provides scientific basis for the implementation of train speed limit operation, traffic control and other plans on rainy days.

Q. Hu, *Natural Disaster Warning System for High-Speed Railway Safety Operation*, Advances in High-speed Rail Technology, https://doi.org/10.1007/978-981-99-7115-2_5

Fig. 5.1 HSR track affected by the heavy rain

5.1 Study Status of the Rainstorm Warning System

The HSR rainstorm monitoring system collects the rainfall information, transmits the collected data to the unified platform and then processes the data to create warning information. This information serves as the data foundation flood control and emergency rescue, track maintenance patrol, traffic scheduling, etc. At present, a lot of application research has been conducted both at home and abroad.

5.1.1 Early Warning System for Flooding Disasters in Japan

Japan's HSR generally implements the rainfall warning operation management system, which establishes warning levels based on the hourly rainfall depth and the cumulative rainfall. The rainfall limits are mainly based on the statistical data of historical disasters and supplemented by several studies on the connection between rainfalls and the integrity of the roadbed, forming a comprehensive system of management that is comparatively scientific. Since its rainfall limits differ for different lines and sections, Japan has installed rain gauges in rainfall-prone areas such as road cutting, fill and tunnel entrances and exits along the lines. Figure 5.2 shows the composition of the rainfall alarm system in Japan.

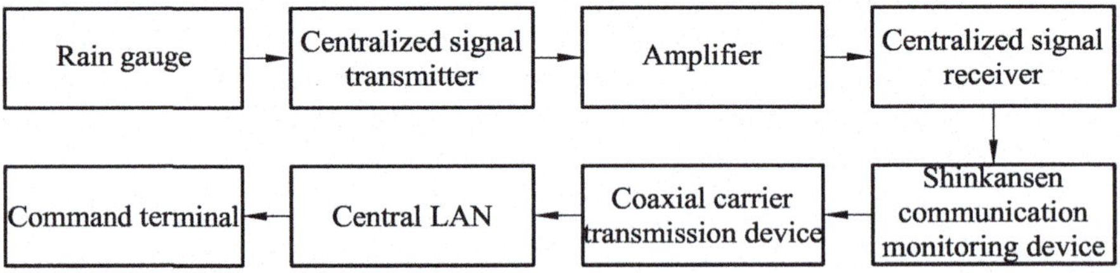

Fig. 5.2 Japan Shinkansen rainfall alarm system

The alarm will sound in the work section and regional dispatching office when the hourly and continuous rainfall surpasses the specified value. The operational control regarding concentrated rainstorms can be divided into speed control and stopping control within the designated interval and mileage. Japan's rainfall warning standards and the train operation control measures are shown in Tables 5.1 and 5.2.

Some HSR line accidents, such as collapse and landslide, cannot be predicted only by monitoring the hourly and continuous rainfalls. The reason is that the continuous rain for a long time makes the groundwater level in the cohesive soil embankment with

Table 5.1 Japan Tokaido Shinkansen rainfall warning standard and train operation control measures (mm)

Run control		Continuous rainfall (24 h gauge)		Hourly rainfall	Continuous rainfall + hourly rainfall	Rainfall report	Note
Alert	Type 3	100–110		25	100 + 20	1 times/h	
	Type 2	120–130		30	110 + 20	2 times/h	Inspect every 3–4 h
	Type 1	> 140		35	120 + 25		Inspect every 2 h
Running speed limit	170 km/h	Area B	/	40	140 + 30 or 160 + 20	2 times/h	Real-time ground inspection with additional inspection if needed
		Area A	/	45	150 + 30 or 180 + 20		
	70 km/h	Area B	/	45	150 + 32 or 180 + 20		
Stop running	General interval	/		50	150 + 40	6 times/h	Emergency inspection for Area B during continuous rainfall
	Viaduct, ballastless bridge	/		70	150 + 60		

Note ① The third type of warning: It refers to the regular inspection and warning in the predetermined zone and the place designated to pay attention on the equipment maintenance; ② In contrast to the third type of warning objects, the second type of warning refers to the routine inspection and warning in the area of geotechnical structures and tunnel entrances; ③ The first type of warning: It refers to the routine inspection and notification in the predesignated area outside the second type of warning object or in the location that is considered to be susceptible to being affected; ④ Warning: The rainfall reaches the promulgated standard, but there is basically no possibility of disaster, and some precursors of disaster can be predicted and should be warned; ⑤ Speed limit operation: The rainfall reaches the promulgated standard, but experience shows that there is no occurrence of disaster. No abnormal rainfall has occured, and there is a possibility of minor disaster; speed limit operation should be considered; ⑥ Stop operation: The rainfall reaches the promulgated standard and there is a possibility of disaster; the operation needs to be stopped; ⑦ Area B: Patrol inspection intervals with continuous rainfall of more than 150 mm and an hourly rainfall of 40 mm; the others are Area A; ⑧ The inspection area is referred to as "the place to pay attention to" when the hourly rainfall hits 50 mm

Table 5.2 Rules for other lines in Japan to stop the train service due to rain

Line area	Conditions for stopping the train
Tokaido Shinkansen	(1) When the hourly rainfall reaches 50 mm; (2) Continuous rainfall can reach 150 mm when the hourly rainfall is up to 40 mm; (3) When stopping train service is considered necessary based on the report of the road maintenance staff or for other reasons
Tohoku Shinkansen	(1) When the hourly rainfall reaches 50 mm; (2) Continuous rainfall of more than 190 mm, hourly rainfall of 40 mm, or continuous rainfall of more than 250 mm, hourly rainfall of 20 mm; (3) Continuous rainfall up to 350 mm; (4) When stopping train service is considered necessary based on the road maintenance staff report or for other reasons
Tohoku Shinkansen	(1) When the hourly rainfall reaches more than 50 ~ 60 mm; (2) Continuous rainfall of more than 200 mm, hourly rainfall of 50 mm; (3) When the continuous rainfall reaches more than 250 mm; (4) When the water level under the bridge beam reaches the stoppage water level; (5) It is considered that the upstream floating objects could seriously erode the bridge or compromise its safety; (6) When stopping train service is considered necessary for other reasons

strong water retention rise unnaturally, leading to the instability of the embankment and foundation. The track slips with the destruction of the base. In light of this circumstance, Japan has introduced a new indicator in addition to the previous hourly and continuous rainfalls: cumulative rainfall, which is continuous or intermittent rainfall within 48 h.

The warning criteria of the cumulative rainfall: For areas having a history of damage from rainfall, 90% of the minimum rainfall at the time of damage is taken as the alarm standard value; for areas without a history of damage, 90% of the maximum cumulative rainfall over the previous 10 years from May to November is taken as the alarm standard value.

5.1.2 Early Warning Systems for Rainfall Disasters in Various Areas of China

The local rainfall monitoring and control will be affected to some extent because of China's vast territory, significant differences in natural circumstances and features of uneven rainfall and diverse rainfall. The overall rainfall monitoring system of China's HSR is shown in Fig. 5.3. The core component of the HSR rainfall monitoring system is the rain gauge. In domestic rainfall monitoring, the tipping rain gauge and siphon rain gauge are mostly deployed by meteorological departments. But at present, the integrated rain gauge is used for rainfall monitoring internationally. The integrated rain gauge uses the acoustical principle of non-mechanical structure to measure precipitation. By detecting the impact force of individual raindrops, it generates the

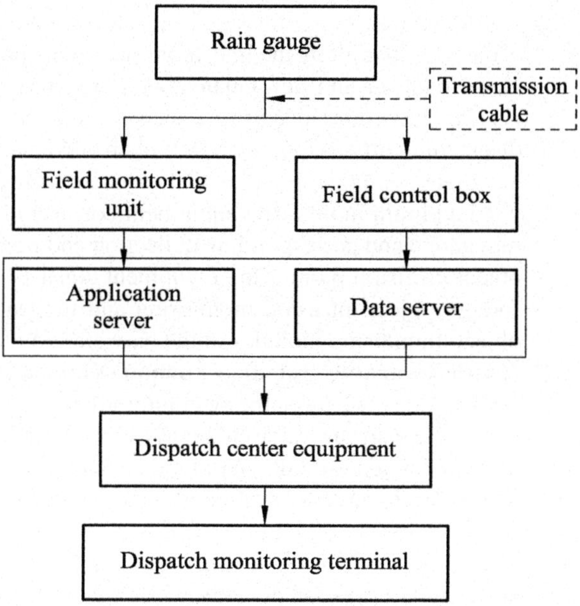

Fig. 5.3 HSR rainfall early warning system in China

signal proportional to the size of raindrops, which is then added up and converted into the total amount of precipitation.

(1) **The rainfall warning system of Shanghai–Nanjing HSR**. The Shanghai–Nanjing high-speed railway in China uses sensors with integrated rain and wind monitoring features to establish flood control technology standards, and the vast majority of monitoring points are set at the same site. The monitoring method of Shanghai–Nanjing HSR is divided into two categories: the hourly rainfall and "24-h rainfall + hourly rainfall." The "hourly rainfall" and "24-h rainfall + hourly rainfall" alarm thresholds are set according to the impact of rainfall on the infrastructure. The Shanghai–Nanjing high-speed rail's rainfall warning system contains an alert and monitoring function. When the rainfall alarm occurs, China Telecom Corporation receives the rainfall alarm information through the disaster prevention and communication server and pops up a prompt, and the train dispatcher checks the disaster prevention terminal to confirm the rainfall alarm and then issues the temporary speed limit order for the corresponding train by the dispatching telephone in time; With the help of China Telecom Corporation terminals and temporary speed limit operation terminals, the train dispatcher can simultaneously set and cancel the temporary speed limit to make the train operate at the automatic speed limit.

(2) **The rainfall warning system of Wuhan–Guangzhou HSR**. China's Wuhan–Guangzhou HSR disaster prevention system is an integrated system consisting of the wind monitoring subsystem, rain monitoring subsystem, foreign object

invasion monitoring subsystem, etc. On-site monitoring units installed at GSM-H base stations along the line, monitoring data processing equipment in Wuhan, New Changsha and New Guangzhou stations, engineering terminals of the engineering duty room in Wuhan, Changsha and Guangzhou integrated work areas, dispatching office equipment in Wuhan and Guangzhou, transmission networks and other components make up the system from the front end to the center. The wind and rain monitoring equipment consists of wind speed and anemometer, rain gauge and corresponding collection and transmission devices. The foreign object intrusion monitoring equipment consists of the dual-grid sensors of the foreign object intrusion monitoring and the trackside controllers and foreign object monitoring modules in the base station.

(3) **Rainfall warning system of Fuzhou–Xiamen HSR**. China's Fuzhou–Xiamen HSR defines rules for the operation control as given in Table 5.3 and determines the warning value by hourly rainfall, daily rainfall, continuous rainfall, or accumulated rainfall, among other factors.

(4) **The rainfall warning system of Nanning Railway**. The first new regulation of flood prevention for China's HSR was introduced by Nanning Railway Bureau at the end of June 2014. This new regulation serves as a guarantee for the HSR safety operation in Guangxi, as well as a reference for flood prevention of the HSR in hilly areas across the country. The *Assessment Standard for Flood Control Sites of Railway with a Speed of 200 km/h and Above* clearly stipulates the rainfall warning value and driving speed of the HSR. The regulations clearly indicate that the train shall operate at a speed limit of 80 km/h or less after emergency repair and opening in case of such disasters as line emptying, bridge and culvert washing away and large debris flow burying the line. The on-site guard must be dispatched if it is difficult to increase or restore the normal running speed in a short time; in the event of mountain landslide, water washing line, river bank scouring and other disasters affecting the subgrade safety, the train shall operate at a speed limit of 160 km/h or less after emergency repairs and opening. It is necessary to send the on-site guard if it is difficult to increase or restore the normal driving speed in a short time. Nanning Railway Bureau additionally specified the rainfall warning value and running speed for key sections of high-speed railway, as shown in Table 5.4.

Table 5.3 Operation regulations of Fuzhou–Xiamen high-speed railway under different rainfall conditions (rainfall unit: mm)

Train speed (km/h)	Hourly rainfall	Daily rainfall	Continuous rainfall
Normal operation	≤ 5	≤ 35	≤ 60
≤ 60	> 5	> 35	> 60
(0,60)	> 5	–	> 82

Table 5.4 Rainfall warning value and driving speed in HSR key sections of Nanning Railway Bureau

Rainfall (mm)	Speed limit (km/h)
$\geq$ 45 within an hour	120
$\geq$ 60 within an hour	45
$\geq$ 75 within an hour or $\geq$ 65 within an hour and $\geq$ 120 continuously	Blocking line

5.2 Early Warning Mechanism for the HSR Safety Operation Under Disasters Caused by Rainstorms

In order to successfully monitor the regional rainfall and ensure the HSR safety operation in the region, we must master the regional climate laws, particularly the laws governing rainfall, due to the regularity of climate characteristics in a certain area.

5.2.1 Spatial Distribution Characteristics of Disasters Caused by Rainstorms

This chapter analyzes the distribution of rainfall in China over the past 30 years and summarizes the spatial distribution pattern of annual average precipitation, especially the characteristics of rainstorm distribution in each area of China, in order to determine the deployment and location of the HSR rainstorm monitoring system.

(1) The distribution characteristics of rainstorm in each area of China. The distribution of heavy rainfall in each area of China gradually decreases from the southeast coast to the northwest inland (such as the direction from Guangzhou to Urumqi), and the precipitation contour shows an obvious northeast–southwest direction. In particular, southern China experiences the highest average precipitation, generally exceeding 1600 mm. The average annual precipitation in Yangjiang, Shangchuan Island and Taishan in Guangdong, Qinzhou and Dongxing in Guangxi and Qiongzhong and Qionghai in Hainan exceeds 2000 mm.
(2) The annual average number of rainstorm days in China. The number of rainstorm days per year in China is consistent with the annual average precipitation distribution, which decreases from southeast to northwest. The distribution of the frequency line of heavy rainfall also basically points a northeast–southwest direction. The number of rainstorm days is the highest in South China and the lowest in northwest China, while some mountainous stations are the least relative to their surroundings.
(3) The initial month of rainstorm in China. The initial month of rainstorm in China can be summarized as "early in the south and late in the north," while the end

month of rainstorm can be described as "early in the north and late in the south." This is consistent with how China's main rain belts move forward and backward. The analysis of the diurnal characteristics of rainstorm shows that the frequency of nighttime rainstorms is lower than that of daytime rainstorms in most areas of South China. On the contrary, the frequency of nighttime rainstorms is larger than that of daytime rainstorms in the eastern part of southwest China, including the Sichuan Basin, Chongqing, Yunnan, Guizhou Plateau and other regions.

5.2.2 The Layout Principle of Rainfall Monitoring Points

In order to ensure the HSR safety operation, rainfall monitoring points should be set up in areas with annual precipitation greater than 200 mm. In principle, the HSR rainfall monitoring points shall be placed in HSR subgrade sections. If possible, the HSR rainfall monitoring point and the wind speed and wind direction indicator should be installed on the catenary pole at the same site. Rainfall, landform and geological characteristics along the line must all be taken into account when determining the environment.

(1) **Distribution principle of the HSR rainfall monitoring points**. The distribution principle of the HSR rainfall monitoring points is mainly based on the analysis of existing rainfall statistics, topography and geological parameters.

① **According to the rainfall statistics**. The point design can be completed in accordance with the meteorological department's current rainfall information.

When the annual precipitation is more than 2000 mm: In the section with the annual precipitation more than 2000 mm, the HSR rainfall monitoring points are spaced 10–15 km apart; in the section where the annual precipitation is more than 1500 mm and less than 2000 mm, the HSR rainfall monitoring points are spaced 15–20 km apart;

When the annual precipitation is less than 1500 mm: In the section with the annual precipitation less than 1500 mm, the HSR rainfall monitoring points are spaced 20–25 km apart.

When the annual precipitation is seriously distributed unevenly: In the section where the annual precipitation is seriously distributed unevenly, the HSR rainfall monitoring points can be appropriately added according to the rainfall data.

② **According to the topography and landform information**. The point design can be carried out based on the topography and geomorphological features of the HSR line area.

The distribution principle of rainfall monitoring points in subgrade section of the HSR: For the ballast track, the rainfall monitoring points of the HSR shall be placed along the ballast track at intervals of 15–20 km; For the ballastless track line, rainfall monitoring points of the HSR shall be placed at intervals of 20 ~ 25 km;

The distribution principle of rainfall monitoring points in the high embankment (cutting) section of the HSR: The rainfall monitoring points in the section with embankment (cutting) higher than 30 m shall be placed at an interval of 10 ~ 15 km; rainfall monitoring points in the section with embankment (cutting) higher than 10 m and lower than 30 m shall be placed at an interval of 15–20 km; the rainfall monitoring points in the section where the height of the embankment (cutting) of the HSR is lower than 10 m shall be placed at an interval of 20 ~ 25 km;

The distribution principle of rainfall monitoring points at the tunnel entrance with a protective net: Rainfall monitoring points shall be set according to the actual needs of the site, with no limit on the number;

The distribution principle of rainfall monitoring points of the drainage section: Because the drainage portion is small, it is possible to install the rainfall monitoring points where heavy rain may lead to poor drainage;

③ **According to the geological circumstance**. In sections with poor geology, such as expansive soil and soft soil subgrade, rainfall monitoring points shall be placed at an interval of 10 ~ 15 km and for other sections shall be placed at an interval of 15 ~ 25 km.

The rainstorm alarm standards and measures shall be determined in accordance with the comprehensive analysis of basic conditions, climate and geographical conditions as well as disaster intensity of the HSR line.

(2) **Layout principle of HSR rainfall stations**. The layout of rainfall stations should be added based on the principle of rainfall distribution, in accordance with the typical spatial and temporal scale characteristics of China's weather system theory, with reference to the design concept of the global ground observation network and the layout principle of the National Meteorological Administration's regional automatic meteorological stations.

Layout principle of densified rainfall stations. Stations can be appropriately encrypted mountainous areas prone to rainfall, road cuttings, frequent flash flood areas with population density.

Layout principle of representative rainfall stations. When installing automatic rainfall monitoring stations, priority should be given to small and medium-sized watersheds in mountainous areas, and stations should be installed in representative sections such as the basin center and rainstorm center as far as practicable. Attention shall be paid to avoiding minefields.

Layout principle of the topographic rainfall station. The rainfall in mountainous areas is affected by the elevation of terrain, so the function of terrain variables shall be fully considered when setting up automatic rainfall stations. The layout can be increased appropriately in areas where mountain torrents pose a higher threat.

Layout principle of the practical rainfall station. The operation management and maintenance conditions such as communication and traffic shall be fully considered when the HSR rain gauge is set up.

Layout principle of the cost-effective rainfall station. The monitoring and early warning platform should incorporate the current rainstorm data from hydrology, meteorology and other departments at various levels (such as 25, 50, 100, 150 mm) in accordance with the maxim of making the most of available resources.

5.3 Influence Mechanisms of the Rainfall on the HSR Safety Operation

Rainstorm monitoring technology is the core technology in the HSR safety storm and disaster prevention. Through the scientific analysis of the rainstorm observation data of all levels in the basic meteorological stations along the HSR line, it is possible to grasp the type and characteristics of the rainstorm in each mile of the HSR line and determine the main focus of rainstorm monitoring in various areas and sections. The rain sensor may be used for monitor the rainstorm and confirm whether the rainstorm reaches the maximum standard before high-speed train operation, so as to control the operation of multiple units in rainstorm weather. The HSR rainstorm warning system will be activated when the maximum precipitation of a high-speed railway section reaches 2.0 mm in 10 min, which can effectively assure the safety operation in case of rainstorm and flood. Therefore, the HSR's technology for preventing and controlling natural disasters heavily relies on storm monitoring system.

5.3.1 Key Parameters of the Rainstorm Warning

In order to systematically and scientifically study the safe operation measures of the HSR under heavy rainfall conditions, this chapter takes the critical rainfall and warning rainfall as the primary operational factors for criteria for evaluating the impact of rainfall on the HSR safety operation.

(1) **Critical rate of rainfall**. The critical rate of rainfall refers to the threshold value of the debris flow disaster caused by heavy rainfall along the HSR. When the rainfall is lower than this value, the possibility of water damage event is very small; when the rainfall exceeds this value, there is a greater chance of a damage event caused by flood. The long-term research findings of domestic and foreign scholars show that the debris flow, landslide and other disasters are closely related to the regional environmental background and precipitation conditions.

The critical rate of rainfall is the prerequisite for the occurrence of debris flows, landslides and other natural phenomena and is also the fundamental benchmark for assessing the hazard level of rainfall to the line safety. Rainfall is one of the necessary factors for the occurrence of disasters. A rainfall of critical rate is the lowest amount of precipitation required to trigger a disaster. The ground environmental conditions are subject to dynamic changes, so the critical rainfall value is a variable value that changes with the ground environmental circumstances within a certain range.

The rainfall of critical rate is composed of previous precipitation and instant rainfall intensity. For the convenience of the on-site operation and management, the daily continuous precipitation is used as the preprecipitation indicator, and the 10 min and 1-h rain intensity are the instant rain intensity. The precipitation directly affects the water content of the soil, thus affecting the critical rainfall rate of water damage. Regarding the time scale, many research use different standards, primarily 15 d, 10 d, 3 d, 1 d, etc. The instant rainfall intensity is what triggers a water damage event, and different studies use different time frames to measure this intensity, mainly 24 h, 12 h, 1 h, 0.5 h, 10 min, etc.

(2) **Rainfall warning**. The rainfall warning refers to the rainfall index developed by the railway department to issue warning signals of water disasters along the line and determine the warning state of the line. It is one of the technical standards for the implementation of the rainfall warning system.

The rainfall warning is the technical application standard created according to the application needs on the basis of the critical rainfall and considering the influence of other factors. The rainfall of critical rate is the basis and foundation for the establishment of rainfall warning system, which is the engineering application of critical rainfall.

The warning rainfall corresponds to the critical rainfall in the composition and also includes two parts: preprecipitation and instant rainfall intensity. Rainfall values are set by the railway departments in accordance with flood control standards, and the specific value is based on where rainfall observation points are located. This chapter divides the rainfall warning values into twelve categories:

The alert principle of the HSR: 10 min' alert, 1-h alert, 3 h' alert, 6 h' alert, 12 h' alert, 24 h' alert and one continuous rainfall alert;

The emergency alert principle of the HSR: 1-h emergency alert, 3 h' emergency alert, 6 h' emergency alert, 12 h' emergency alert, 24 h' emergency alert and one continuous rainfall emergency alert.

5.3.2 Rainfall Warning Mechanism

Railway departments use the HSR rainfall alert system as a safety guarantee system in order to direct flood control dispatching and assure the traffic safety of the line during flood season. The rainfall warning system of the HSR consists of three parts, namely, warning standard, defense standard and management system. It is a warning system that is put in place by level, section and division of labor and joint control.

The separation of alert sections along the HSR, the defining of precipitation circumstances entering various alert states and the formation of alert standards for

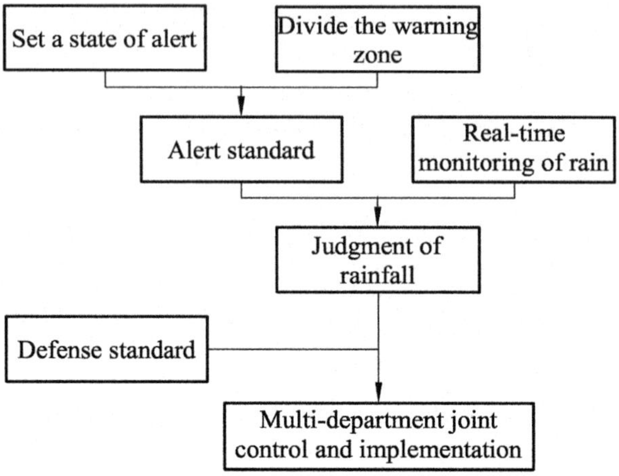

Fig. 5.4 Process of creating and implementing the rainfall warning system

classification and section are the key components of the construction of the HSR rainfall alert standards (see Fig. 5.4).

(1) **Operation Mechanism**. The HSR rainfall warning system is powered by rainfall monitoring. The full-time permanent rain guage is equipped along the high-speed rail line. When precipitation happens, the rain gauge will determine the alert status of the line in the section based on the amount of rainfall that was really measured and will then send alert information to each department. After receiving the warning information, all HSR departments must immediately implement the necessary warning measures. The HSR track maintenance department, for instance, will immediately dispatch personnel to patrol the line when the rain gauge issues the "Attention" warning signal; the train and maintenance departments must also notify the train and maintenance personnel in the section to operate the train at a speed not exceeding 60 km/h and strengthen the train lookout measures (see Fig. 5.5).
(2) **Management Measures**. Throughout the flood season, at any moment, the HSR line is continuously in regular operation without rain, normal operation with rain, "attention" alert operation, "critical" alert operation and "blocked section." The five states, the boundary measurements of each of the railway functional department under each operational state are displayed in Table 5.5, each of which is controlled by one of the five states.

5.3.3 *Warning Thresholds for the Risk of Rain*

The division of early warning sections along the HSR, the definition of precipitation conditions entering various early warning states and the formation of the threshold standards for hierarchical and sectional early warning are the main components of threshold standards for the HSR rainfall disasters.

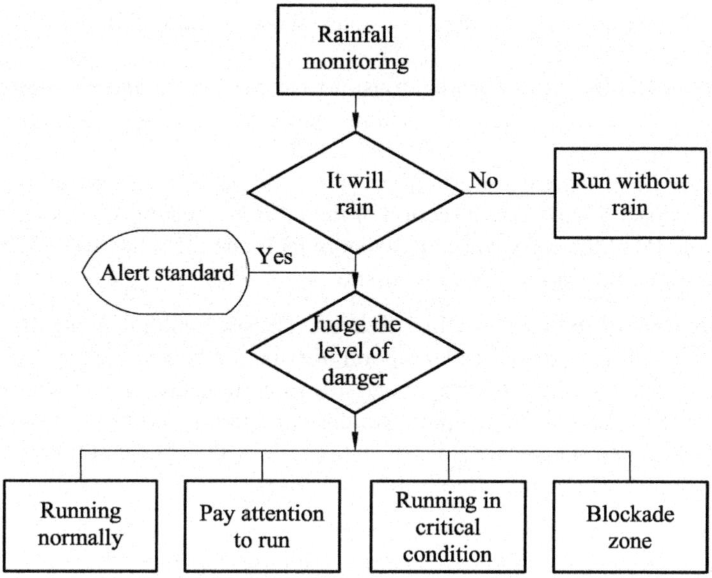

Fig. 5.5 Operational flowchart of the rainfall warning system

Table 5.5 List of alert measures of railway functional departments for each running state

Running state	Deterrent measures	
	Works' department	Mechanical and vehicle services' department
Normal operation without rain	Track patrol	Pay attention to lookout
Normal operation with rain	Track patrol	Pay attention to lookout
Pay attention to alert operation	Assign personnel to patrol in the rain until the alarm is withdrawn	Run no faster than 60 km/h, and bolster the train watch
Emergency alert operation	Increase the number of people on patrol until the alert is lifted	Run at a speed of no more than 40 km/h, and strengthen the train lookout
Blocking interval	The entire staff kept watch in the rain until the alarm was removed	Block the line until the obstruction is lifted

(1) **Division of the HSR warning sections**. The characteristics of the rain conditions along the HSR serve as a guide for determining the threshold value for early warning, so the division of the early warning section should fully consider these characteristics and also consider the ground conditions, line conditions, on-site operation and management. The following guidelines should be followed in the process of zoning:

Principle I of the early warning division: follow the rainfall division in the flood season along the HSR;

Principle II of the early warning division: The geological and geomorphological conditions and geological hazards within each segment of the HSR are relatively consistent;

Principle III of the early warning division: The HSR line conditions and flood resistance capacity within each section of the HSR are generally consistent;

Principle IV of the early warning division: facilitate the setup and administration of the local rainfall warning system on-site.

(2) **Indicators of the HSR section division**. The precipitation along the HSR line exhibits intricate spatial and temporal distribution characteristics. Therefore, it is required to choose the proper classification indicators and methodologies to measure the spatial and temporal similarity of precipitation at stations along the HSR. The two statistical values of stations along the HSR that follow have been chosen as the indicators for section division.

Indicator I of the HSR section division: precipitation during the flood season, which serves as a general indicator for precipitation statistics.

The daily precipitation in the flood season being greater than 30 mm is indicator II of the HSR section division. The daily precipitation ranges from 30 to 50 mm at the stations along the HSR that are situated in a hilly region, with an average of more than 96% rainstorm. Therefore, using daily precipitation throughout the flood season for the statistics of rainstorm along the HSR has additional significance.

5.3.4 Three Levels of Warning for Rainfall Disasters

The HSR rainfall catastrophe warning is now divided into "attention warning," "emergency warning" and "blocking section" of the flood warning. The two more significant of the three levels of warning status are "attention warning" and "blocking warning."

The "attention warning" state determines whether the patrol staff will venture outside in the rain and consequently whether the staff will identify and broadcast the danger signal of high-speed railway line in time when the water disaster occurs.

Whether the HSR transportation is interrupted depends on the "blocking section" status.

The "emergency warning" state is the intermediate state between the two. In comparison to the earlier, it raises the early warning degree of HSR lines. In contrast to the latter, it is necessary to make corresponding technical preparations for HSR lines to enter the "blocking section" status. The procedure for establishing alert standards is shown in Fig. 5.6.

In the three-level warning state of the HSR rainfall disaster, the warning rainfall is determined by the cooperation between the necessity for engineering applications and crucial precipitation conditions water damage. First, to ensure that early warning

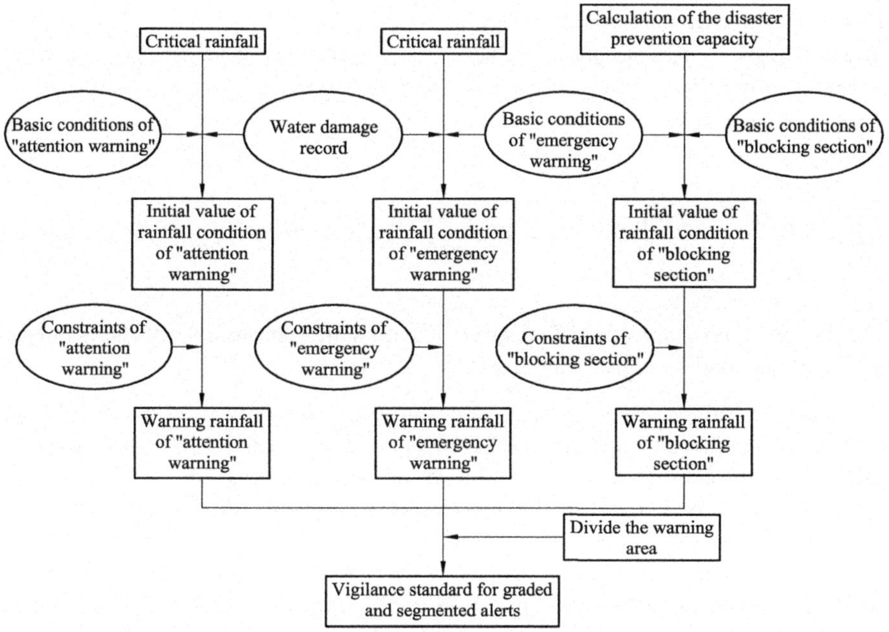

Fig. 5.6 Flowchart of vigilance standard development

systems at all levels can promptly detect and prevent water disasters, and to guarantee the safety of HSR traffic during flood seasons; Second, the early warning status at all levels should not be too sensitive. Otherwise, even while it ensures the absolute safety of the HSR operation, it will substantially disrupt the continuation of the HSR transportation and cause more interference with the normal operation of the HSR; finally, it should be practical for on-site operation. In a word, early warning rainfall should adhere to a technical standard that is highly reliable, causes little disruption to transportation and is highly operable. As a result, the following fundamental prerequisites and restrictions are established for the HSR line to enter the three-level early warning state:

① Basic requirements for entering the state of "attention warning:" the calamity has not yet occurred or has essentially not occurred, and there is some level of advance. Constraints of this state: The number of patrols for public works during the rainy season in a particular region shall be controlled within 5 ~ 10 times;
② Basic prerequisites for entering the status of "emergency warning:" the danger begins to unfold, and there is almost no advance. Constraints of this state: The number of trains moving slowly due to a downpour in a single stretch shall be kept to a maximum of 3 times and a minimum of 1 time;
③ Basic prerequisites for reaching the state of "blocking section:" The water disaster has already caused the line damage; the disastrous heavy rainfall is taking place, and the continuous rainfall is sufficient and continues to rain heavily; disasters

Table 5.6 Critical rainfall at alert level three

Guard interval Dominant factors	Pay attention to the alert	The speed limit warning	The blockade alert
Daily continuous rainfall (mm)	50 ~ 100	101 ~ 150	151 ~ 200
1-h rainfall (mm)	8 ~ 15	16 ~ 30	31 ~ 40
10-min rainfall (mm)	2 ~ 4	5 ~ 6	7 ~ 8

may occur on a large scale. Constraints of this state: The number of single interval blocking shall be kept within the range of 0.1–1.

There are differences in the measuring criteria which are used to establish the fundamental conditions for high-speed railway lines to reach the "attention warning" and "blocking section" states. The concept of "attention warning" is based on the potential for water emergencies. Therefore, its early warning rainfall value is based on the minimum critical rainfall value of the documented water disasters as a reference and thus represents the requirement of a specific amount in advance. "Blocking section" is measured by whether flood disaster endangers HSR lines. Therefore, its early warning rainfall value is mostly determined by evaluating the disaster resistance ability of the line, and its value is limited by the number of block intervals, as indicated in Table 5.6.

5.4 Rainstorm Warning System for the HSR Safety Operation

The HSR rainfall monitoring system is a subsystem of the disaster prevention and safety monitoring system, which enables real-time monitoring of rainfall data along the HSR. The HSR rainfall warning system's unified processing platform consists of components such as the on-site monitoring equipment (rain gauge, field control box, transmission cable), on-site monitoring unit, railroad bureau central system, dispatching center equipment, engineering dispatching equipment and corresponding transmission network equipment. The HSR disaster prevention system is built on a communication and transmission system that incorporates data gathering, storage, analysis and processing and other integrated intelligent monitoring systems, etc. It is a part of the system for operations and dispatch.

(1) **Rain gauge of the HSR early warning system**. Rain gauge, often known as the rain sensor, is a crucial tool for tracking the precipitation. The precision of the rain gauge directly affects the accuracy of the rainfall measurement and the reliability of the rainfall record report. Therefore, the accuracy of the rain gauge plays a vital role in the timely issuance of decision instructions for the HSR flood control and disaster relief to safeguard the safety of lives and property.

The rain gauge of the HSR early warning system is connected to the on-site monitoring unit via the signal cable, and the on-site monitoring unit is set up in the communication base station or station machine room near the rain gauge to complete the collection, preliminary analysis and processing of rainfall data.

(2) **Monitoring unit of the HSR early warning system**. The monitoring unit transmits data and information to the central system of railroad bureau via the transmission network, completes the analysis and processing of rainfall data, generates the information about monitoring, warning and alarm as well as the operation management suggestions. It then transmits this information to the appropriate monitoring and reshowing terminals of transfer station and engineering department.

(3) **Dispatching management center of the HSR early warning system**. The railway dispatcher can set and remove the temporary speed limit with the aid of CTC terminal and temporary speed limit operation terminal, causing the trains to automatically travel at the speed limit; additionally, the relevant train can get the command to temporarily reduce speed in the event of heavy rain.

In due time by the wireless transmission function of the dispatching telephone and the CTC system

(4) **Track division management center of the HSR early warning system**. The department of works management sets up the patrol and rescue plan according to various alert information. The power supply of the rain gauge is centralized by the UPS in the closest machine room using the signal cable. The equipment in the machine room is centrally powered by UPS. High-speed railway sections have a large number of GSM-R base stations, which may serve as the system's foundation thanks to their flexible rain gauge distribution, short-wired access distance, safe and reliable access mode and faultless features.

Figure 5.7 depicts the flow diagram for the HSR rainfall warning system. When the rainfall intensity reaches the threshold value, the rainfall monitoring emergency management system is initiated. The workflow of the emergency system is shown in Fig. 5.8.

In the HSR rainfall warning system, the threshold used for the determination is the real-time dynamic threshold value, which is produced by clustering based on the past disaster statistics and real-time monitoring data. The emergency measures of rainfall are divided into three levels.

(1) **Warning interval for the HSR safety operation**. The early warning intervals in the HSR rainfall early warning system are classified according to rainfall.

Third-level warning: When the continuous rainfall reaches 100–110 mm, the communication module will report the rainfall every hour;

Second-level warning: When the continuous rainfall reaches 120–130 mm, the communication module will report the rainfall every 0.5 h, and the staff will examine every 3–4 h;

First-level warning: If the continuous rainfall is greater than 140 mm, the communication module will report the rainfall every 0.5 h, and the staff will examine every 2 h.

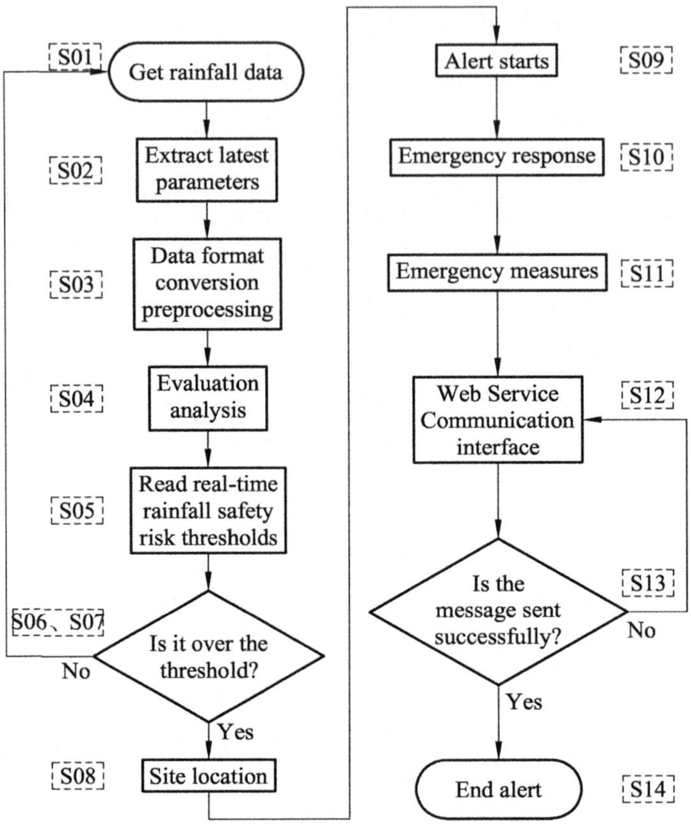

Fig. 5.7 Workflow of the rainfall monitoring and early warning system

(2) **Speed limit threshold for the HSR safety operation**. The HSR rainfall early warning system limits the speed of high-speed trains according to rainfall.

About 200 km/h is the third-level speed restriction barrier. The running speed of high-speed train shall not be higher than 200 km/h;

Second-level speed limit threshold: 160 km/h. The operating speed of high-speed train is limited to 160 km/h;

First-level speed limit threshold: 60 km/h. A high-speed train shall not operate at a speed greater than 60 km/h.

(3) **Stop operation measures for the HSR safety operation**. It is required to cease the operation in segments according to general sections, viaducts, ballastless bridges, etc.

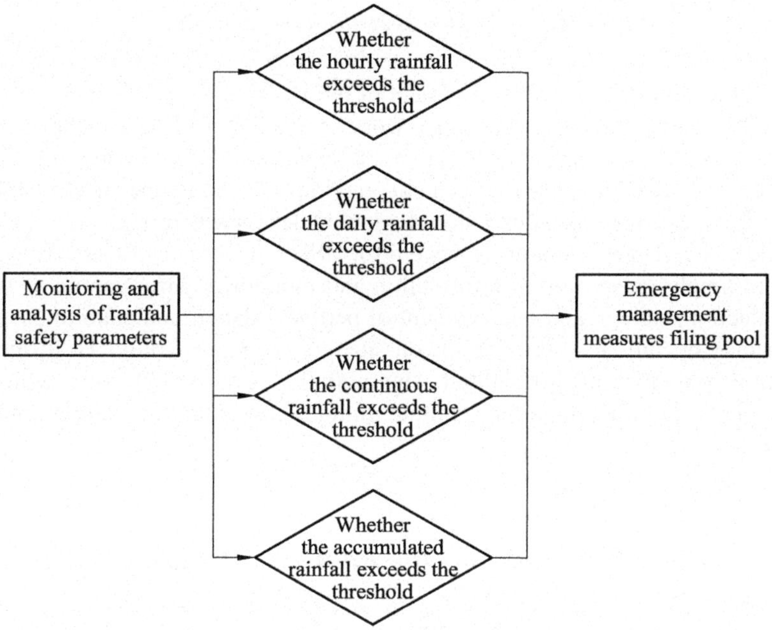

Fig. 5.8 Workflow of the rainfall monitoring emergency management system

5.5 Early Warning System for Rainfall Catastrophe Safety Operations

Regional flood disasters in the rainy season are prone to cause the collapse of the HSR roadbed, especially debris flows, landslides, dangerous rock falls and other natural hazards in difficult mountainous areas, which pose a great threat to the HSR safety operation. The purpose of the HSR heavy rainfall disaster monitoring and forecasting system is to provide alarm information for dispatching and maintenance management and effectively prevent the impact of the rainfall disaster on the safety of line and train operation by timely monitoring the rainfall of the HSR along the line, especially the rainfall of the roadbed and special line section, and timely understanding the dangerous situation such as the roadbed settlement and collapse along the HSR in the rainy season. Therefore, from the perspective of engineering design, this chapter proposes the overall architecture, system function, system equipment configuration, monitoring information transmission and other schemes of the HSR rainfall monitoring system in order to serve as a guide for the design of the HSR rainstorm monitoring and forecasting system.

5.5.1 Warning Content of the Rainstorm Disaster

The existing on-site detection equipment of the HSR rainstorm detection system is often only installed in the maintenance work area and at a small number of stations along the line. The monitoring of subgrade settlement, collapse and other threats along the line in the flood season is hampered in particular by the lack of collection points in some dangerous mountainous regions and parts with problematic geological subgrade. A real-time monitoring system for HSR rainstorm disasters is required in order to timely grasp rainfall information and monitoring information, as well as the collection, storage and management of pertinent data to maintain to ensure the HSR safety operation. At present, the on-site monitoring data of the HSR rainstorm detection system are mainly rainfall. The rainfall indicators concerned by the HSR flood control department are: ten-minute rainfall, hourly rainfall, daily rainfall and continuous rainfall.

5.5.2 Data-Aware Model for the Rainstorm Disaster Warning

The data processing model of the HSR storm disaster warning is mainly investigated by using the Paulhus formula. A precipitation process with two variables: precipitation R and total precipitation ephemeris T, is provided. The sample outcome (value I) turns into a random variable if the precipitation intensity I is taken from T in any sufficiently small hourly interval. In terms of meteorology, the likelihood of a precipitation intensity I (instantaneous value I) sampled from T is correlated with the distribution of the precipitation intensity in time over the course of the period T. Using the relationship between relative distribution function and meteorological entropy, it can be concluded that the distribution function of precipitation intensity in time is the same function as the probability distribution of I at random sampling.

(1) **Rainfall defining equation for the HSR**. The HSR rainstorm warning system is the variation law of the rain intensity when the precipitation intensity I_0 surpasses $\bar{I} - I_0$ exceeds a $I - I_0$ certain $\bar{I}$ specific lower limit I, so $I \geq I_0$, can be used to replace in place of the average rain intensity and rain intensity, respectively. Therefore, the rainfall defining equation for the HSR is

$$f(I) = (\bar{I} - I_0)^{-1} \exp\left[-\left(I_0 - \bar{I}\right)\left(\bar{I} - I_0\right)\right], \tag{5.1}$$

where $f(I)$ is the probability density of the precipitation intensity I derived from the maximum entropy in the precipitation duration T;

$\bar{I}$ is mean precipitation intensity (mathematical expectation).

(2) **Rainfall time equation for the HSR**. Definition: r is the total amount of precipitation in the period when the intensity of the precipitation exceeds I, then the

function relationship is

$$r = \int_{I}^{\infty} ITf(I)dI, \tag{5.2}$$

$$t = T\exp\left[-\left(I_0 - \bar{I}\right)\left(\bar{I} - I_0\right)\right], I \geq I_0. \tag{5.3}$$

Rainfall time equations for the HSR is

$$\frac{r}{R} = \frac{t}{T} \cdot \left(1 - \left(1 - I_0 \cdot \bar{I}\right) \ln \frac{t}{T}\right), \tag{5.4}$$

where

$\frac{r}{R}$ is the relative precipitation amount;
$\frac{t}{T}$ is the corresponding relative precipitation ephemeris.

In order to streamline the calculating process and accelerate the warning speed, let $I_0 = 0$, then the rainfall time of the HSR Eq. (5.4) becomes

$$\frac{r}{R} = \frac{t}{T} \cdot \left(1 - \ln \frac{t}{T}\right), \tag{5.5}$$

where

T is the maintenance time of a rainstorm in a certain area (storm calendar time);
R is the total amount of rainstorm during time period T, and t is a discrete period inside of T. Therefore, $t \leq T$;
R is the maximum amount of rainfall produced during period t.

5.5.3 A Warning System for Rainstorm Disasters

In order to efficiently monitoring the rainstorm and assure the HSR safety operation, building a rainstorm monitoring and early warning system is important for the HSR safety operation. The secondary architecture of monitoring center system and on-site detection equipment is used by the HSR rainstorm monitoring and forecasting system. This architecture is made up of the monitoring center, on-site detection equipment, monitoring terminal, transmission network and power supply and other components. Among them, the transmission network can be wired or wireless, and the system architecture is presented in Fig. 5.9.

As depicted in Fig. 5.9, the composition of the HSR rainstorm warning system is as follows:

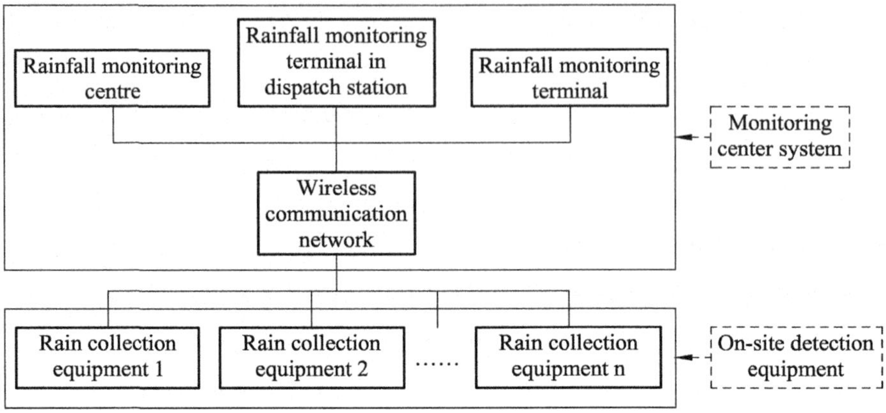

Fig. 5.9 HSR rainstorm warning system

① Monitoring center of the HSR rainstorm warning system. It receives the rainfall monitoring data and equipment functioning status uploaded by the on-site equipment, summarizes, analyzes and processes the data collected in accordance with the protocol, creates monitoring and warning information and sends it to the relevant monitoring terminal. It has the statistical analysis function of rainfall monitoring information according to the specified time period and provides the management staff with the query and display of the monitoring alarm and equipment fault light information, and the report printing of the real-time monitoring and centralized monitoring management shall be carried out for the HSR central system, on-site monitoring equipment and network. Interfaces for communication with national meteorological services are set aside to receive information on disaster prediction, early warning and other topics;

② On-site rainfall monitoring equipment of the HSR rainstorm warning system. The HSR on-site rainfall monitoring system collects, analyzes, processes and uploads the rainfall data while continuously monitoring the information along the line in real time;

③ Monitoring terminal equipment of the HSR rainstorm warning system. The HSR monitoring terminal equipment performs early warning disposal while displaying rainfall information and operation management recommendations in the form of graphics, text, alarm, etc. There are five tiers to the HSR rainfall monitoring and early warning system. Each level of the early warning system comes with its corresponding speed limit value, as well as capabilities for information querying and reporting. The classification of monitoring and early warning is shown in Table 5.7;

④ HSR transmission network and power supply system. The HSR transmission network completes the trustworthy transmission of rainfall monitoring data and early warning information among the on-site monitoring equipment, monitoring terminal equipment and monitoring center system. All components of the rainfall

Table 5.7 Warning level of the HSR rainstorm disaster

Early warning level	Hourly rainfall (mm)	Daily rainfall (mm)	Continuous rainfall (mm)	Speed limit (km/h)
Level 1	[0,25)	[0,90)	[0,100)	180
Level 2	[25,35)	[90,105)	[100,130)	140
Level 3	[35,45)	[105,130)	[130,160)	100
Level 4	[45,55)	[130,155)	[160,190)	60
Level 5	> 55	> 155	> 190	0

monitoring system have reliable working power thanks to the power supply system.

5.5.3.1 Early Warning Architecture for Rainstorm Disasters

The HSR rainstorm disaster monitoring and early warning system (see Fig. 5.10) includes the control module, rainstorm parameter information collection module, storage module, evaluation and early warning module and communication module. The rain parameter collection module, storage module, evaluation and warning module and communication module are each connected to the control module in turn. The fundamental module utilized to regulate the operations of the aforementioned modules is the control module. The evaluation and warning module processes and analyzes the data on the parameters of the precipitation that have been collected, and it can communicate evaluation and warning data about the precipitation.

The HSR rainstorm parameter information collection module is deployed every 30 km along the HSR line and is strengthened in the areas that are vulnerable to disasters caused by rainstorm, such as cutting, filling, tunnel entrance and exit, in order to monitor the rainstorm information at the location of the device and extract the spatial coordinate information. It can also record the monitoring information and spatial coordinate information in the storage module.

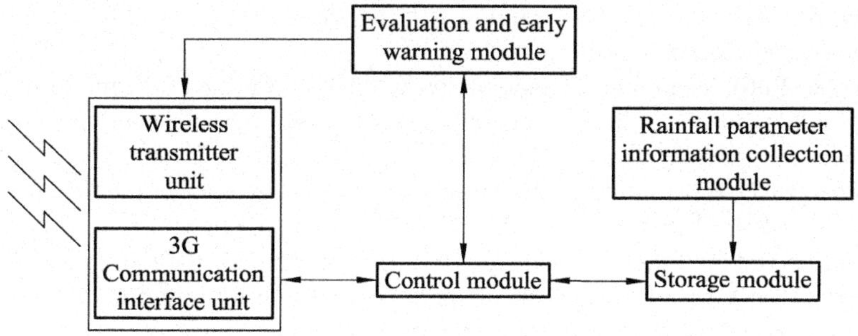

Fig. 5.10 Structure of the HSR rainstorm information collection and warning system

Only monitoring the hourly and continuous rainfalls cannot reliably predict some HSR line mishaps, such as collapse, landslide. The cause is the prolonged, non-stop rain, which causes the groundwater level in the cohesive soil embankment with strong water retention to rise abnormally, making the embankment and foundation to become unstable. Sliding then happens, and the base is destroyed. In light of this circumstance, it is necessary to organize and keep track of the accumulated rainfall in special sections, such as continuous or intermittent rainfall within 48 h; the storage module is used to store aforementioned rainfall parameter information and can transmit the rainfall parameter information through the communication module under the control of the control module.

The information about rainfall parameters that were gathered above is evaluated and warned about using the HSR evaluation and warning module. The evaluation results are used to determine the safety level of the impact of the current rainstorm on the HSR, and the information for the corresponding HSR rainfall emergency measures is obtained using the safety level standard of the HSR rainfall. Under the control of the control module, it can convey the evaluation information of the rainstorm safety level and information about HSR rainfall emergency measures through the communication module, in order to carry out the warning process of the HSR rainstorm safety, as shown in Fig. 5.11.

The following steps are involved in the processing and analysis of the rainfall parameter data from the HSR rainstorm disaster monitoring and early warning device.

Step 1 (S01): The monitoring module for rainfall parameter information gathers data about the location of the device that is relevant to the rainfall;

Step 2 (S02): The control module retrieves the most recent rainfall parameter information in accordance with the current time;

Step 3 (S03): The control module translates the information from the obtained data into AD;

Step 4 (S04): The control module contacts the evaluation and early warning module to conduct the analysis of the pretreatment rainfall parameters. Calculating the average value of the current rainfall parameters and the fluctuation range of the recently collected rainfall parameters are the specific objectives of the analysis of the rainfall parameter;

Step 5 (S05): The real-time rainfall risk threshold data derived by clustering is read by the control module;

Step 6 (S06): The control module checks the data on the real-time rainfall risk threshold with the rainfall parameters fluctuation interval generated in Step 4 above;

Step 7 (S07): If the rainfall safety information does not go over the risk threshold, return to Step 2 and examine the rainfall parameter information again. Step 8 should be taken if the rainfall safety information exceeds the risk threshold;

Step 8 (S08): Identify the built-in spatial coordinate information of the rainfall control module to find the affected high-speed railway lines;

Step 9 (S09): After positioning, the control module initiated the alert command and calls the alert process in the evaluation and warning module;

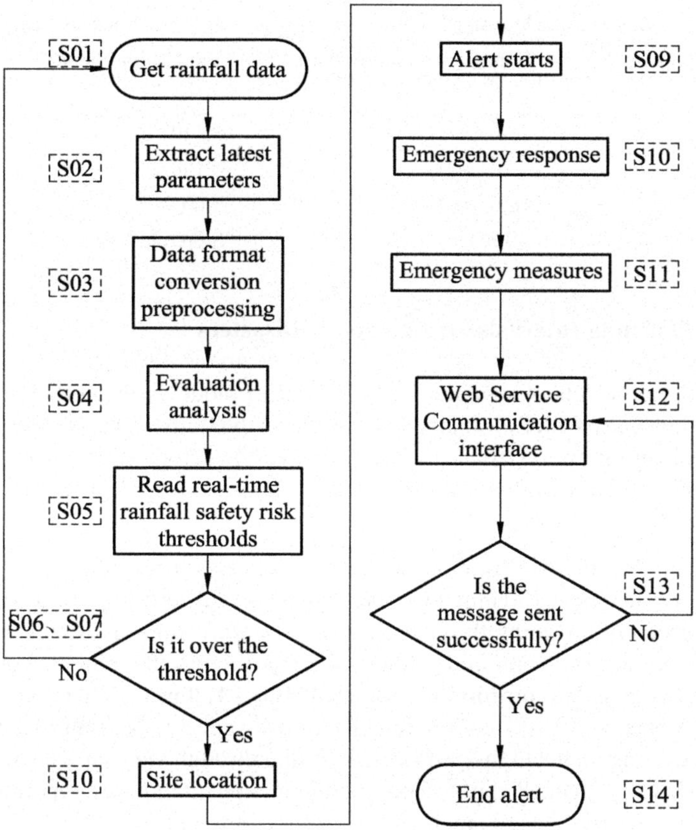

Fig. 5.11 Warning process of the HSR safety under rainstorm disasters

Step 10 (S10): Respond to emergencies in accordance with the early warning guidance;
Step 11 (S11): Determine the corresponding emergency filing level and measures of high-speed railway line based on the corresponding rainfall safety level of the HSR line obtained from the corresponding rainfall safety assessment. As part of the emergency procedures, passengers are evacuated to a safe location after being warned, and their operations are slowed down or stopped;
Step 12 (S12): Send the text-based emergency response information using a mobile communication interface or wireless transmission module;
Step 13 (S13): Verify that the text message containing the emergency action is sent successfully: if so, proceed to Step 14; if not, restart the communication module to send the text information with the emergency action information;
Step 14 (S14): Exit alert.

The corresponding control mode of the HSR for various levels of rainstorm disaster environment is displayed in Table 5.8.

Table 5.8 Various warning levels and vehicle control strategies in conditions of heavy rain

Early warning level	Continuous rainfall (mm)	Vehicle control mode (km/h)
Level 1	< 100	< 180
Level 2	100 ~ 130	< 140
Level 3	130 ~ 160	< 100
Level 4	160 ~ 190	< 60
Level 5	> 190	Stop running

5.5.3.2 Warning Intervals for Rainstorm Disasters

Under different levels of rainfall, the HSR control mode varies, too. The real-time dynamic threshold, which is obtained by clustering the statistical data of past disasters and the real-time monitored data, is the threshold utilized for evaluation and judgment in the HSR rainfall disaster warning system. There are five degrees of emergency procedures for rainstorms:

① **Warning level I: continuous rainfall less than 100 mm**. When the continuous rainfall is less than 100 mm, the railway agency will send the first warning to the high-speed trains going through the area. The train control speed is lower than 180 km/h, and the communication module will report the rainfall every hour;

② **Warning level II: continuous rainfall of 100 ~ 130 mm**. When the continuous rainfall is between 100 and 130 mm, the railway agency will transmit a two-level early warning to the high-speed trains passing through the area. The train control speed is lower than 180 km/h. The communication module will report the rainfall every 0.5 h and the personnel will inspect it every 2 h;

③ **Warning level III: continuous rainfall of 130 ~ 160 mm**. When the continuous rainfall reaches 130 ~ 160 mm, the railway department will send a three-level early warning to the high-speed trains going through the area. The train control speed is lower than 100 km/h. The communication module will notify the rainfall every 0.5 h and the personnel will check it every hour;

④ **Warning level IV: continuous rainfall of 160 ~ 190 mm**. When the continuous rainfall hits 160 ~ 190 mm, the railway agency will send a four-level early warning to the high-speed trains passing through the area. The train control speed is lower than 60 km/h. Every 15 min, the communication module will notify the rainfall and the staff members will check it every 30 min;

⑤ **Warning level V: continuous rainfall exceeds 190 mm**. When the amount of continuous rainfall exceeds 190 mm, the railway department gives a five-level warning to the high-speed trains passing through the area. At that point, the trains cease running and the area is closed and trains are not allowed to enter.

5.6 Summary of This Chapter

The HSR rainstorm early warning technology is an important part of the HSR safety precautions and disaster avoidance countermeasures. The HSR rainfall disaster monitoring and early warning system is built in this chapter, focusing on the characteristics of HSR and merging them with the rainfall characteristics in China.

Due to China's enormous size, natural conditions vary widely, the amount of rainfall is quite unequal and rainfall characteristics are also very diverse. These factors will affect local rainfall monitoring and control to some extent. Therefore, according to diverse local rainfall characteristics, this chapter proposes a targeted rainstorm monitoring system and early warning process by examining the mechanism of rainstorm influence on the HSR and the distribution characteristics of China's multiple rainy places. The monitoring and early warning system of the HSR rainfall disaster constructed in this chapter increases the accuracy of the rainstorm monitoring in China.

Bibliography

1. Yi L (2006) Research on railway remote rainfall monitoring and flood control dispatching system. Central South University, Changsha
2. Zhang M (2009) Research on pilot project of coastal railway disaster prevention (wind, rain) safety monitoring system. Railway Sign Commun Eng 6(5):36–38
3. Deng H (2013) Discussion on the scheme of railway rainfall monitoring system in southwest mountainous area. Commun Inf Technol 7(5):100–102
4. Hu Q, Bian L, Tan M (2020) Data perception model for safe operation of high-speed rail in rainstorm environment. Transp Res Part D Transp Environ 83:102326
5. Falahati A (2021) Improve safety and security of intelligent railway transportation system based on balise using machine learning algorithm and fuzzy system. Int J Intell Transp Syst Res 19(3):274–286
6. Tang J, Yao L, Jiang L, Hua M (2002) Spatial statistical characteristics of precipitation in flooding season on Cheng-Kun railway (north section) and the demarcation of rainfall alarm zones. China Railway Sci 23(6):95–99

Chapter 6
A Ecological Warning System for the HSR Safety Operation

Locating in the intersection area of the circus Pacific tectonic zones, where the two active tectonic zones converge, is the fundamental reason for the wide variety of geological disasters in the country to be blame for. According to statistical data, there are nearly 30,000 natural disasters, such as collapse, landslide and debris flow, in China on average annually, including certain special and major disasters. For example, on January 18, 2010, a major geological disaster-induced collapse of Guangzhou–Nanning HSR in Baiyun Tunnel led to five deaths.

Types of geological hazards also vary. According to the nature and location of the geological-induced disasters, there are 48 kinds of common geological disasters in 12 categories in total. Among them, collapse, landslide and debris flow, as the main types of geological disasters, are characterized by strong outburst, wide distribution and concealment and often take a heavy toll on huge economic development and lives every year. Statistical data in recent years show that the number of geological disasters in China remains quite stable, as the natural environment and geographical conditions of that year are the major reason behind. Geological disasters are beyond human control but still predictable and preventable.

The impact of geological hazards on the HSR is increasingly prominent along with its rapid development in China. China's HSR lines are often under direct face threat of multiple geological hazards as they traverse various natural areas. At present, there are more than 10,000 large debris flow gullies, about 1000 large and medium-sized landslides, more than 1000 collapses, more than 4000 serious collapses along the railway lines in China. In recent years, although with the strengthening of early warning systems in China, the hazards of debris flow have been greatly reduced in numbers and prevented, the accidents occurred have still caused disasters. For example, in June 2011, the heavy rainfall along the Baiguo to Puxiong section of Chengdu–Kunming line induced 17 debris flows. Fortunately, the timely warning did not lead to a traffic safety accident of debris flow, but the driving was interrupted for more than 100 h, causing significant economic losses. In May 2014, a debris

Q. Hu, *Natural Disaster Warning System for High-Speed Railway Safety Operation*, Advances in High-speed Rail Technology, https://doi.org/10.1007/978-981-99-7115-2_6

Fig. 6.1 Trains suspended on the Guangzhou–Shenzhen line after debris flow

flow in the section from Shenzhen North to Guangming City of the Guangzhou–Shenzhen line resulted in the suspension of all HSR services from Guangzhou South to Shenzhen North (see Fig. 6.1). The accident raised significant economic losses in spite of no casualty.

The speed of the HSR, which is generally 200 ~ 320 km/h, is faster than that of the ordinary railway, which means that geological disasters on the journey such as debris flow will lead to extremely serious transportation interruption, loss of people's lives and properties and great disturbance to the construction, maintenance and normal operation of the HSR as well. Therefore, in order to guarantee the HSR safety operation, it is necessary to study the early warning method for the HSR safety operation under geological hazards.

A typical debris flow can be divided into three sections, formation area, circulation area and deposition area (Fig. 6.2). Intense heavy rainfall is the main trigger for the occurrence of debris flow. Rainstorm can also trigger flash floods that lead to a series of natural disasters such as collapse, landslide and debris flow, seriously threatening the HSR safety operation. That is why railway departments around the world have been extremely careful about geological hazards along the lines by actively carrying out geological hazard defense, reducing the hazards of natural disasters through regional hazard surveys and adopting measures such as line bypassing and engineering prevention. Against these backdrops, this chapter focuses on the early warning management of the HSR safety operation under geological hazards.

6.1 Studying on the Geological Hazard Warning System

As geological hazards mainly refer to debris flow in this book, the impact of debris flow hazards on the HSR safety operation is discussed in this chapter. Today, in order to prevent the irreversible damage of debris flow to the HSR, each HSR system in the world is equipped with corresponding debris flow disaster warning systems. While rainfall is the main triggering factor of mountain disasters, it is still impossible for

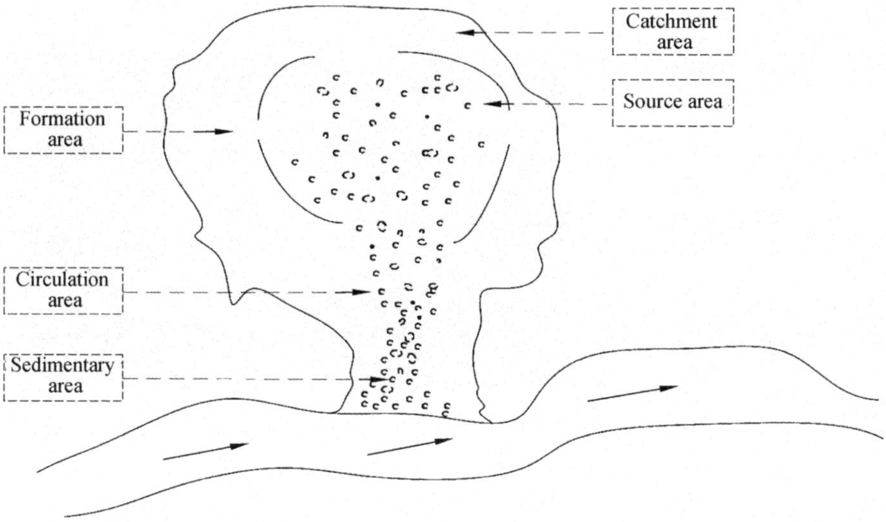

Fig. 6.2 Debris flow

human beings to completely be free from its impact by solely relying on engineering measures for huge randomness of natural disasters in terms of the outbreak location and scale. Even if possible, the required astronomically high investment in disaster prevention and control often makes the social and economic benefits little cost-effective. As disaster prevention measures such as flood prevention and early warning, with small investment and high social and economic benefits, is particularly effective in mountainous debris flow disaster prevention, the debris flow disaster prediction and early warning system thus has become an important means to ensure the safe transportation of the HSR.

6.1.1 Debris Flow Disaster Warning System in Switzerland

Located at the northern foot of the Alps, the debris flow gully in the east of Switzerland is a representative project of the kind. The drainage area of the gully is 1.8 km^2. Since 1834, the debris flow has broken out every 5–10 years and even was as high as six times once in one year. With the frequent occurrence of extreme climate, the frequency of the debris flow increases, bringing the total volume of solid materials washed out by each debris flow ranges from $0.02 \times 10^4\,m^3$ to $7.0 \times 10^4\,m^3$. The Swiss Federal Government therefore has installed a set of debris flow monitoring and early warning system in the debris flow gully (Fig. 6.3).

The Swiss debris flow monitoring and early warning system includes a rainfall monitor, four sets of earth sound monitors, two sets of ultrasonic mud level monitors and two sets of infrared camera devices, in which two sets of camera are equipped

Fig. 6.3 Debris flow monitoring and warning system in Switzerland

with special recording and image transmission equipment, while other instruments are connected to specially developed data acquisition and data processing terminals. The data acquisition terminals process the debris flow disaster information received by the sensors in real time and then instantly transmit it to the monitoring center of the Water Resources and Geology Office of Swiss Federal Government. The monitoring and early warning system is fully automated and unattended. Apart from alerting debris flow hazards in target areas, the system has another core function, which is to observing and collect information on debris flow formation and movement for scientific research on debris flow disaster prevention and mitigation. It can be seen that the early warning system for debris flow disasters in Switzerland is both advanced in performance and comprehensive in function and operable.

6.1.2 Debris Flow Disaster Warning System in Austria

Austria uses a contact system to sense the movement and arrival of the debris flow and send back information so as to provide early warning of debris flows along the railway. But using sensing line for early warning is not suitable for the gully section where the erosion and deposition of debris flow change greatly. With the erosion of gully bed deepens in the section with large scouring, the mud level of debris flow cannot reach the position of the original metal sensing line, which means that the sensing line loses the function of the alarm and misses reports. The impact force in the deposition section of debris flow gully is not very damaging as the velocity of debris flow there is low. Reports will also miss when the sensing line is buried by debris flow but not broken, and the alarm function is lost. Impact force detection and early warning are another method of the kind. In this situation, an impact force detection sensor in the gully will be placed where the debris flow passes and sends back signals if it detects the impact force of the debris flow on the sensor when the debris flow passes. The early warning of debris flow disasters is realized as the impact

Fig. 6.4 Sensing line in the debris flow gully

force, speed and mud level of the debris flow can all be estimated, thus. Figure 6.4 shows the sensing line deployed in the debris flow gully in Austria.

Currently, the statistical model mainly used in the flow early warning is based on the rainfall monitoring information with a single indicator value. But the disconnection between the flow warning system and the real-time disaster also caused low accuracy of debris flow disaster early warning and prediction, large fuzzy prediction interval and high possibility of false and missing reports. Therefore, it is imperative to improve the key physical parameters of the debris flow warning and the accuracy of early warning. The debris flows generally occur in mountainous areas, but the harsh environment along the HSR and consequential poor communication and traffic conditions make it very difficult to obtain information on the precursors, imminent disasters and disasters. That is why there is an urgent need for practical debris flow disaster warning technology.

6.2 Influence Mechanisms of the Debris Flow on the HSR Safety Operation

Debris flow disaster often leads to serious consequences in a wide range, as it happens mostly at night all of sudden with strong force. Moreover, in addition to the huge direct economic losses, the disaster also takes a heavy indirect toll on economy and society. Therefore, it is of great significance to systematically analyze the impact mechanism of debris flow to ensure the HSR safety operation and reduce the damage caused by debris flow.

6.2.1 Characterization of Debris Flow Hazards

The main types of geological hazards along the HSR line are landslide, collapse, debris flow and unstable slope. For example:

① Landslide is a more serious type of geological hazard that is widely distributed but often occurs along the HSR line and nearby areas;

② Along the HSR and its surrounding areas, collapse is one of the main geological hazards. Collapses are widely distributed, numerous in quantity, relatively small in scale, and often occur suddenly, making them difficult to avoid and frequently resulting in casualties and property losses;

③ Debris flow is a form of concentrated transport of loose solid accumulation materials in the basin during the process of water and soil loss in mountain areas. The rainfall acting on the ground environmental system destabilizes the loose solids on the surface, gets them involved water movement under the action of their own gravity and hydrodynamic force and sends debris flow to channels when the mass of solids containing exceeds a certain limit.

(1) **The four main damage mode of the debris flow to the HSR**: impact, siltation, blockage and scouring.

Hazard mode 1: impact: When the debris flow moves at a high speed, the debris flow directly impacts the HSR buildings, causing a strong collision between them, and destroys the buildings when the impact force is greater than the bearing capacity;

Hazard mode 2: siltation: After the debris flows out of the mountain pass, the sediment gradually accumulates in the piedmont area of the open terrain and gentle gully bed and even bury the line, station, bridge and culvert and tunnel, causing the railway interruption if the elevation of the HSR line passing through the area is low;

Hazard mode 3: blockage: The HSR will be blocked by debris flow when it passes through the debris flow gully with bridges and culverts that have small flow capacity or improperly connected longitudinal slopes above and below the bridge. If the debris flow blocks bridges and culverts, it will directly wash the subgrade, endangering the line and other facilities;

Hazard mode 4: scouring: Debris flow is characterized by large scouring and silting. The gully bed can be scarred as deep as 15 m after a debris flow. It not only erodes the gully bed, but also hollows out the beam, abutment, and revetment foundation, or cuts through the top of the tunnel, resulting in damage to buildings that cross the gully and the corresponding protective measures.

(2) **Factors affecting the debris flow**. The debris flows along the HSR are affected by many factors, which can be concluded as two categories. One is basic factors of geological disaster activities including geological structure, landform and geotechnical conditions. Only with certain basic geological conditions, can landslide activities occur; the other is triggering factors consisting of the earthquake, rainstorm and human activities. Under certain landform and geotechnical conditions, earthquake, rainstorm and some human engineering activities may also

trigger geological disasters. However, we can only determine the inducing factor as the dominant one, as any geological disasters are triggered by the joint action of many factors.

(3) **Impact of the debris flow on traffic**. At present, HSR traffic accidents caused by debris flow often happen when the debris flow leader destroys the train running on the bridge or in the culvert or when the train breaks into a place that has been damaged or buried by debris flow. In addition, the types of disasters prone to occur in different locations of the HSR are also different. The HSR generally consists of the subgrade section, bridge section, tunnel section, elevated section and station, among which geological disasters are more likely to occur in tunnel and bridge section. Rainstorm type of the debris flow gully, the main debris flow disasters along the HSR in mountainous areas, is still in exploration due to the complexities in the development process, formation process, circulation process and disaster causing mode of rainstorm type of the debris flow. Many theoretical studies are in a semi-qualitative and semi-quantitative state.

6.2.2 Key Parameters for Debris Flow Disaster Warning

The common debris flow disaster along HSR in mountainous areas is mainly storm-type gully debris flow, which is also a common geological disaster along mountainous railways in the rainy season. Debris flow is one of the evolution processes of the natural environmental in small watershed after the destabilization of slope and gully bed accumulation under the action of gravity and hydrodynamics, but it can cause serious disasters. Due to the complexities in the development process, formation process, movement process and disaster mechanism, rainstorm-type debris flow is still in the exploration with semi-qualitative and semi-quantitative state studies being the most. However, for a specific debris flow gully, there are always landmark parameters (see Fig. 6.5) in the process of formation, movement and disaster indicating that a debris flow is about to form or has formed. These landmark parameters can be used for warning the debris flow disaster and reducing debris flow disaster losses and social impacts.

(1) **Rainfall**. Rainfall has many impacts on the formation of debris flow as it is the most active part in the process:

① Rainfall provides water components for the formation of debris flow;
② Rainfall provides hydrodynamic conditions for the confluence of loose solid materials and promotes the potential energy of loose solid materials into kinetic energy;
③ Rainfall speeds up the instability and destruction of loose solid materials on the slope surface of the source area of debris flow, which later are collected into the gully bed to participate in debris flow under the action of rain washing and erosion;

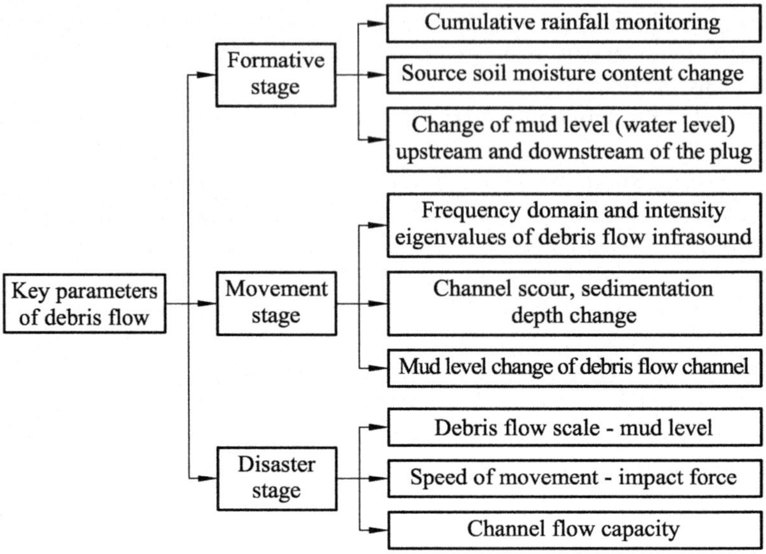

Fig. 6.5 Key parameters of the debris flow disaster warning

④ After the rainwater converges into the gully, the increase of flow, speed, scouring and carrying capacity are the main hydrodynamic conditions for the restart of gully bed materials.

(2) **Infiltration depth of the rainfall**. The biggest difference in the occurrence of debris flow is the participation degree of various types of solid debris material in the debris flow basin. There are mainly three sources of object material in the debris flow fluid:

① The objects from the slope of the debris flow source and the soil mass of the bank slope under the erosion of the surface water and the intrusion of the surface water are unstable and often enter the gully bed after mixing with the gully flow to participate in the debris flow activities;
② The objects come from the landslide and collapse in the debris flow valley basin, disasters of which are partially or completely involved in the debris flow activities;
③ The objects come from the gully bed material, which starts again when it is scoured by water or dragged by debris flow. At a certain water speed and longitudinal gradient of the gully bed, the gully bed material starts again to participate in debris flow activities under the action of a certain strength of water scouring and undercutting.

The transformation of instable material source (soil mass) into debris flow involves the process of water filling, saturation, liquefaction and movement of the soil mass in the source area of debris flow, which means that the rainfall infiltration depth of the soil mass in the source area of debris flow can be monitored, and the debris flow disaster warning can be conducted through moisture.

(3) **Mud level of the gully**. Another source of material involved in debris flow activities is mainly landslides or bank collapse on both sides of the gully in the basin. The blocking (stopping) water flow barriers or embankments are formed after the landslide starts or bank collapses into the ditch, and large and small weirs are formed in the ditch. Generally, there is only small scale of the debris flow or little possibility for disaster occurrence if there is no gully landslide or bank collapse of solid material to cause the outburst. It can be said that landslide or bank collapse is a decisive factor of the maximum flood flow of the debris flow and one of the important influencing factors of the debris flow disaster. See Fig. 6.6 for details.

In this chapter, according to the mud level characteristics of the "flood storage (energy)-dike breaking and energy release" type gully section, the mud level monitoring instruments shall be deployed in the upstream and downstream sections of potential landslide or blockage, to monitor the rise and fall of the mud level (water level) and then carry out early warning of debris flow disasters.

(4) **Infrasound wave in the process of debris flow movement**. Debris flow is a kind of thick fluid full of sediment and rocks that flows in valleys and gullies at a speed of several meters to tens of meters per second (usually 10 ~ 20 m/s) between high-sediment flow and block (landslide, collapse, etc.). The infrasound signals it sends out have unique characteristics, including frequency,

Fig. 6.6 Schematic diagram of the formation of gully blocking and debris flow

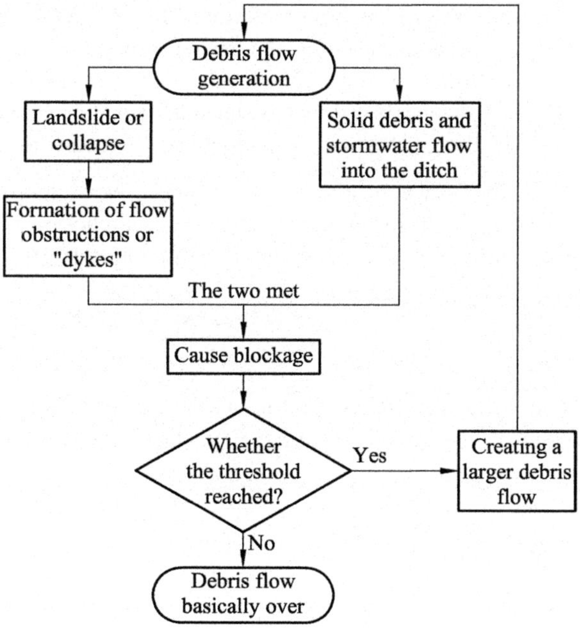

dominant frequency amplitude, and duration. According to the multi-year infrasound signal monitoring data of Jiangjiagou debris flow from Dongchuan Debris flow Observation and Research Station of Chinese Academy of Sciences, the infrasound signal of debris flow is a deterministic signal with a simple harmonic sine wave in determined predominant frequency.

6.2.3 Threshold Discrimination of Geological Hazards

The debris flow that causes hazards to buildings has another landmark parameter, the impact force, whose signal enables the estimation of the impact force size, speed, mud level of the debris flow, which in turn enables debris flow disaster alarm. The technology is to place impact force monitoring sensors in the gully through which the debris flow passes to monitor the impact force of the debris flow on the sensor when the debris flow flows through the sensor. However, as the device is easy to be washed away in large debris flow activities, it is only used for the debris flow observation research at present.

(1) **Criteria for discriminating the scale of geological hazards**. The scale of geological hazards can be divided into four types: giant, large, medium and small. In this chapter, according to the scale of geological hazards, the grade classification of geological hazards in the HSR system is shown in Table 6.1.
(2) **Rainfall statistical model**. At present, many countries adopt rainfall statistical models for the forecasting of the debris flow disaster measured by critical rainfall thresholds of 10-min rainfall, hourly rainfall and 24-h rainfall. For example, since the 1970s, both Japan and China began to study the rainfall parameter with the 10-min rainfall as the basis for the debris flow forecasting as the threshold is believed to be most closely related to debris flow outbreaks. Based on the historical debris flow disaster statistics, this chapter suggests that 10-min rainfall has the greatest relationship with the occurrence of the debris flow.

Chen Jingwu, a scholar from Chinese Academy of Sciences, has achieved good research results through its numerous researches on 10-min critical rainfall for debris flow outbreaks. The discriminant formula for 10-min critical rainfall is as follows:

$$R_{i10} \geq \mu_1 R_{i10} - \mu_2 \text{Pa} \geq \mu_3 R_{i10}, \tag{6.1}$$

Table 6.1 Criteria for the classification of debris flow disaster

Level	Total volume ($10^{4\,\mathrm{m}^3}$)	Number of dead or missing (persons)	Direct economic loss (ten thousand yuan)
Giant	> 50	≥ 30	≥ 1000
Large	20 ~ 50	10 ~ 30	500 ~ 1000
Medium	2 ~ 20	3 ~ 10	100 ~ 500
Small	< 2	< 3	< 100

where

R_{i10} is the 10-min rainfall required for a debris flow;
Pa is the early rainfall;
$\mu_1 R_{i10}$ is the 10-min rainfall required for a debris flow when Pa $\to$ 0;
$\mu_3 R_{i10}$ is the minimum rainfall required for a debris flow when Pa $\to$ Pa*, and Pa* is the rainfall required when the solid material supplying the debris flow reaches saturation;
μ_1, μ_2, μ_3 are coefficient related to many conditions for the formation of the debris flow in the basin, where $\mu_2 = \frac{\mu_1 R_{i10}}{\text{Pa}^*}$.

A functional model of the real-time rainfall monitoring network system along the HSR line. Rainfall is an important triggering factor for the formation of debris flow and thus is of great significance in the HSR debris flow monitoring and early warning system. Given that rainfall is the basis of debris flow early warning, the functional relationship of the real-time rainfall monitoring network system along the HSR is as follows:

$$Y = R \cdot M, \tag{6.2}$$

where

R is rainfall coefficient;
M is discriminant coefficient of surface debris flow gully.

A function model of the real-time debris flow monitoring network system along the HSR. The real-time monitoring network system of rainfall along the HSR is developed according to the discriminant Eq. (6.2). The rainstorm condition function R and the basin environment dynamic function M are established to predict and warn whether debris flow is formed. The rainfall condition function of debris flow is established according to the combined relationship between the action mechanism, magnitude difference and spatial location of various active factors in the two systems:

$$R = K\left(\frac{H_{24\text{h}}}{H_{24\text{h}(D)}} + \frac{H_{1\text{h}}}{H_{1\text{h}(D)}} + \frac{H_{1/6\text{h}}}{H_{1/6\text{h}(D)}}\right), \tag{6.3}$$

where

K is the correction factor of the early rainfall, $K > 1$; when there is no early rainfall, $K = 1$;
$H_{24\text{h}}$ is the 24-h maximum rainfall;
$H_{24\text{h}(D)}$ is the 24-h critical rainfall for the debris flows that may occur in the area;
$H_{1\text{h}}$ is the 1-h maximum rainfall;
$H_{1\text{h}(D)}$ is the 1-h critical rainfall for the debris flows that may occur in the area;
$H_{1/6\text{h}}$ is the 10-min maximum rainfall;
$H_{1/6\text{h}(D)}$ is the 10-min critical rainfall for the debris flows that may occur in the area.

The relationship between the rainfall condition function and debris flow outbreak is shown in Table 6.2.

The critical rainfall of $H_{24h(D)}$, $H_{1h(D)}$ and $H_{1/6h(D)}$ varies between regions and river basin. According to the annual average rainfall zoning of China's rainstorm zoning results, the critical rainfall of debris flows that may occur in all parts of the country is concluded from statistical analysis based on the rainfall observation data of the debris flow observation sites in the region where debris flows have and have not occurred historically (see Table 6.3).

(3) **Threshold value of the infrasonic wave in the process of debris flow movement**. The frequency intensity of infrasound wave, more than 20 dB, is higher than other environmental noise frequency. The value of sound pressure is related to the type and scale of debris flow, where by the thicker the debris flow is, the more rocks are contained, the greater the flow velocity is, and the greater the amplitude and the corresponding increase of sound pressure is.

The properties of infrasound: amplitude of about 0.5 ~ 4 Pa, frequency of 5 ~ 10 Hz (about 6 Hz). For the infrasound of the debris flow, in the low part of the near ground, the absorption coefficient of the infrasound signal is roughly $a = 10 \sim 4\,\mathrm{db/km}$. Infrasound waves can be transmitted at a speed of 344 m/s in the air at room temperature.

The characteristics of infrasound: low frequency of infrasound wave, small absorption of the medium when propagating, far propagation distance of the wave and unimpeded transmission through a small gap in the medium.

As HSR generally runs through remote and harsh mountainous areas, it is difficult to obtain information on debris flow disasters, causing serious man-made and natural damage. According to these characteristics of debris flow infrasound waves mentioned above, a new type of debris flow infrasound acquisition and alarm device can be developed which can even reach tens of minutes in the early warning.

The intensity of the infrasound can be directly affected by the scale of the debris flow. According to the observation and research on the infrasound data of debris flow, the infrasound pressure value of debris flow is approximately proportional to the flow of debris flow. Therefore, the threshold value of the debris flow infrasound pressure can be defined in the HSR system, as shown in Table 6.4.

Table 6.2 Relationship between rainfall condition function and debris flow outbreak

Name		Probability of debris flow
R value range	Early warning threshold	–
< 3.0	–	Safe rainy situation
3.0–4.0	The early warning value of 3.0	Rainfall that could trigger mudslides
4.0–5.0	The early warning value of 4.0	Probability of debris flow outbreak < 0.2
5.0–10	The early warning value of 5.0	Probability of debris flow outbreak < 0.2–0.8
> 10	The early warning value of 10	Probability of debris flow outbreak > 0.8

Table 6.3 Limit values of $H_{24\text{ h}(D)}$, $H_{1\text{ h}(D)}$ and $H_{1/6\text{ h}(D)}$ for possible debris flow

Average annual rainfall zone	$H_{24\text{ h}(D)}$	$H_{1\text{ h}(D)}$	$H_{1/6\text{ h}(D)}$	Representative region
> 1200	100	40	12	Mountainous areas of Zhejiang, Fujian, Guangdong, Guangxi, Jiangxi, Hunan, Hubei, Anhui, suburban Beijing, Liaodong and western Yunnan, southeastern Xizang provinces (autonomous regions)
1200–800	60	20	10	Some mountainous areas of Sichuan, Guizhou, eastern and central Yunnan, southern Shaanxi, eastern Shanxi, Inner Mongolia, Heilongjiang, Jilin, western Liaoning and Northern Hebei provinces (autonomous regions)
800–400	30	15	6	Some mountainous areas in northern Shaanxi, Gansu, Inner Mongolia, Ningxia, Shanxi, Xinjiang, western Sichuan and Xizang provinces (autonomous regions)
< 400	25	12	5	Northwest Yunnan, western Sichuan Hengduan Mountain region, Qinghai, Xinjiang, Tibet and Gansu, Ningxia two provinces and regions west of the Yellow River

Table 6.4 Definition table of infrasound pressure threshold of debris flow

Level	Early warning color	Debris flow discharge (m^3/s)	Infrasound pressure (Pa)
Giant	Red	> 1000	> 10
Large	Orange	800 ~ 1000	8 ~ 10
Middle-large	Yellow	500 ~ 800	5 ~ 8
Medium	Green	300 ~ 500	3 ~ 5
Small	Cyan	100 ~ 300	1 ~ 3
Microminiature	Blue	10 ~ 100	0.1 ~ 1
Miniature	Purple	< 10	< 0.1

From Table 6.4, it is known that the scale of the debris flow can be roughly judged according to the monitored infrasound pressure value of the debris flow.

(4) **Early warning threshold of the mud level**. The catastrophic process of the deterministic debris flow disaster depends on whether the debris flow can safely flow in the debris flow gully or say the relative size of the outburst flow of debris flow and the safe flow of bridges, culverts and gullies which can be converted into mud level information at the fixed channel section. On the selected flow section, the scale of debris flow can be determined directly according to the mud level. Since the mud level monitoring and early warning system of debris flow developed on the basis of the mud level information of real-time monitoring of debris flow can calculate its flow, identify whether the debris flow occurs and send out early warning signals of the debris flow, it is intuitive, easy and can discriminate accurately. Therefore, according to the ratio between the monitored mud level information and the safe overflow mud level information, an early warning level can be issued in real time according to the preset procedure to ensure the HSR safety operation. The discriminant of mud level early warning level of debris flow disaster is as follows:

$$F(\text{danger}) = \frac{H}{h}, \tag{6.4}$$

where

$F(\text{danger})$	is the degree of danger;
H	is the actual mud level of the debris flow;
h	is the safe overflow mud level of bridges, culverts and ditches.

Taking the actual situation in China and the particularity of the HSR into account, the discrimination level of mud level early warning threshold of debris flow is shown in Table 6.5.

(1) **Level 1 warning of the mud level of HSR debris flow disaster, 0 < *H* < normal flood level:** When 0 < *H* < normal flood level, railway bridges, culverts and ditches are in safe working condition, the possibility of disaster is very small. Combined with other factors to issue a warning, the train reports the amount of debris flow every 1 h during normal operation;
(2) **Level 2 warning of the mud level of HSR debris flow disaster, normal flood level < *H* < 0.8 h:** When the normal flood level < *H* < 0.8 h, the HSR bridges, culverts and ditches are in safe working condition, with little possibility of disaster. While prompt early warning has been issued, the train continues to operate normally but reports mud and stone flow every 45 min;
(3) **Level 3 warning of the mud level of HSR debris flow disaster, 0.8 h ≤ *H* < 0.9 h**: When 0.8 h ≤ *H* < 0.9 h, the HSR bridges, culverts and ditches are in a relatively safe working state, with a low probability of disaster formation. As primary early warning is issued, the train reports debris flow every 30 min while continuing normal operation;

Table 6.5 Mud level warning of the debris flow disaster

Early warning level		Actual mud level value of debris flow	Early warning model
Level 1	No warning	$0 < H <$ normal flood level	Operate at normal speed
Level 2	Suggestive warning	Normal flood level $< H < 0.8$ h	Report the debris flow every 45 min
Level 3	Primary early warning	0.8 h $\le H < 0.9$ h	Report the debris flow every 30 min
Level 4	Formative early warning	0.9 h $\le H < 1.1$ h	The train speed is less than 170 km/h
Level 5	Non-disaster warning	1.1 h $\le H < 1.3$ h	The train speed is less than 120 km/h
Level 6	Disaster warning	1.3 h $\le H < 1.5$ h	The train speed is less than 80 km/h
Level 7	High disaster warning	$H \ge 1.5$ h	Stop

(4) **Level 4 warning of the mud level of HSR debris flow disaster, 0.9 h ≤ H < 1.1 h**: When 0.9 h ≤ H < 1.1 h, the HSR bridges, culverts and gullies are in the critical working state of disaster, and debris flow is in the formation state, with possibility of causing disaster, but has not yet reached the disaster state. At this time, a formative early warning of debris flow disaster is issued, and the train runs at a reduced speed of no more than 170 km/h;
(5) **Level 5 warning of the mud level of HSR debris flow disaster, 1.1 h ≤ H < 1.3 h**: When 1.1 h ≤ H < 1.3 h, the HSR bridges, culverts and gullies are in a dangerous working state of disaster, with high possibility of causing disasters. At this time, a non-disaster early warning of debris flow disaster is issued, and the train runs at a reduced speed of no more than 120 km/h;
(6) **Level 6 warning of the mud level of HSR debris flow disaster, 1.3 h ≤ H < 1.5 h**: When 1.3 h ≤ H < 1.5 h, the HSR bridges, culverts and gullies are in a dangerous state of overflow, with high possibility of causing disasters. At this time, a disaster early warning of debris flow hazard is issued, and the train runs at a reduced speed of no more than 80 km/h;
(7) **Level 7 warning of the mud level of HSR debris flow disaster, H ≥ 1.5 h**: When H ≥ 1.5 h, the HSR bridges, culverts, ditches are in a very dangerous state of overflow and have turned into a disaster. At this time, a high-level early warning for debris flow disaster is issued, and high-speed trains are stopped.

Therefore, according to the safe overflow mud level, the mud level warning threshold for the HSR safety operation is set as five thresholds: 0.8 h, 0.9 h, 1.1 h, 1.3 h and 1.5 h.

6.3 Debris Flow Warning Architecture for the HSR Safety Operation

Subsystems consisted the debris flow disaster warning system for the HSR safety operation is defined as follows.

(1) **Rainfall disaster warning system**. The system can predict the probability of its occurrence and issue a warning when the rainfall in a certain section reaches a certain value. The determination of the rainfall threshold is the most critical part in the rainfall warning system and hence should be determined according to different geological formations and the actual situation around.
(2) **Infrasound hazard warning system**. At the moment when the debris flow occurs, special sound waves will be emitted from the source (at a speed of about 344 m/s and takes air as the medium with much higher speed than that of the debris flow and basically same intensity). Once there is a debris flow within a certain range, it will be found by the infrasonic monitor immediately, so as to gain valuable time for avoiding disasters.
(3) **Ultrasonic mud level meter early warning system**. After the debris flow has been assembled, the scale of the debris flow will be judged. The early warning system is composed of ultrasonic mud level meter and is installed at the gully with possible flow passing. If the previous two early warning systems are "misjudged," the mud level meter will send an early warning signal through the system when the debris flow reaches a certain size and passes through the monitoring section of the ultrasonic mud level meter.
(4) **Radar and other early warning systems**. Integrated by computer and modern communication technology, the early warning system can also be connected with the monitoring terminal of the water conservancy department to achieve real-time monitoring, which takes only a few seconds from the rainfall or the sound wave reaching the set value to sending out the signal.

With aforementioned subsystems, the early warning system for the HSR debris flow monitoring is constructed, with early warning process shown in Fig. 6.7.

6.4 HSR Debris Flow Warning System

Made of mountain torrent containing solid materials (mud, sand and stone), a debris flow must has three basic conditions: steep terrain, abundant and loose solid materials and enough water. Often breaking out suddenly with great speed and energy, it has particularly obvious destructives in the mountainous areas of China, especially in the southwest, a region with steep terrain, complex geological structures and active seismic zones. However, structures, earthquakes and mountain construction in recent years and human irrational activities have also provided a large number of solid material sources for the formation of debris flows. The hazards of debris flow to

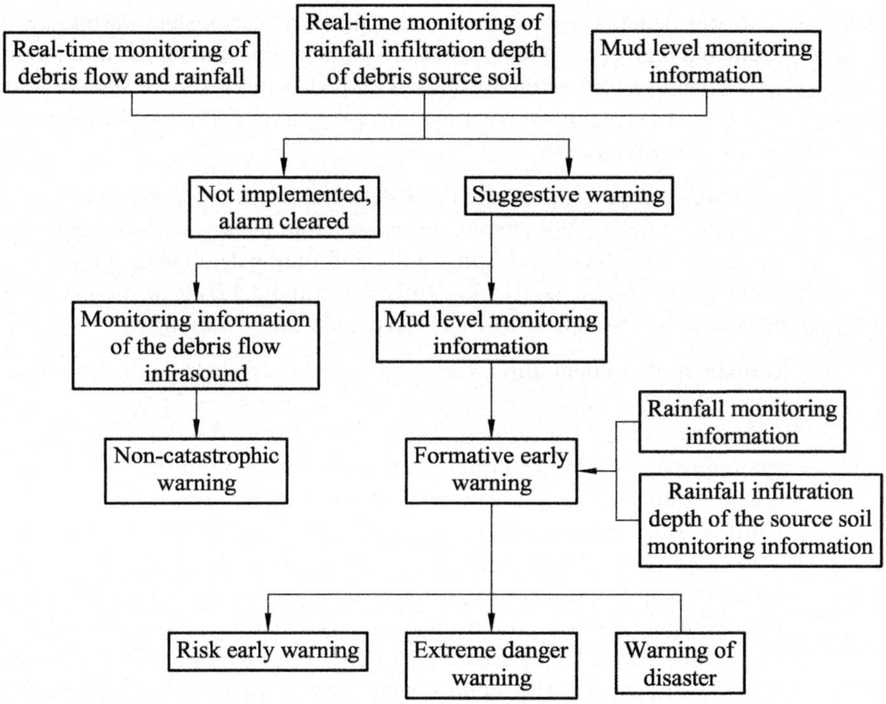

Fig. 6.7 Flowchart of the debris flow warning system

the HSR are mainly shown in the following aspects: siltation, overflow, washout, scouring, abrasion, curve climbing, river blocking and river squeezing.

6.4.1 Early Warning of the Debris Flow Disaster

Abundant precipitation with high intensity is the main cause of the debris flow disaster. Heavy rainfall can pose serious threats to the HSR safety operation as it can trigger flash floods and a sequential series of natural disasters such as debris flows, landslides and collapses. That is why the railway authorities attach great importance to the work related to geological hazards along the line by actively carrying out geological disaster prevention. Based on the regional disaster investigation, many natural disasters have been reduced by means of the line bypass, engineering prevention and control measures. However, due to the intrinsic randomness in the rainfall, the main excitation factor of mountain disasters and the outbreak location and the scale of the rainfall, it is impossible for human beings to completely achieve the purpose of eliminating disasters by only applying engineering measures. Moreover, the extremely high engineering investment of the disaster prevention and control often offset the economic benefits it brings. In contrast, flood forecasting and warning, with low investment and high economic benefits, are more practical in the field of mountainous debris flow disaster prevention. Based on that point, it can be seen that the debris flow disaster forecasting and warning system is an important tool essential to ensure the HSR safety operation.

I. **Early warning parameters of the HSR debris flow**. In order to guarantee the safety operation of HSR, this chapter determines four factors, real-time accumulated rainfall in the debris flow basin, rainfall infiltration depth in the upper body of the debris flow source, debris flow infrasound and debris flow mud level, as key parameters for the debris flow monitoring and early warning.

   (1) **Rainstorm flood peak flow**. The solid sediment carried by rainstorm flood flow in the basin means that the flow of solid debris flow is directly related to the flow of liquid rain flood. Therefore, the peak flow of debris flow is generally calculated by multiplying the explosive rain flow and a correction number related to the characteristic index of debris flow.
   (2) **Rainfall infiltration depth of the source soil**. As an important part of the debris flow source, the stability of the slope soil of the debris flow source is a very complex problem since the stability is closely related to the slope, water content, physical and mechanical properties of the soil of the debris flow source and affected by rainfall infiltration, which is also under the influence of the vegetation condition of the slope, the nature of the soil, the rainfall process curve and other characteristics.
   (3) **Infrasound**. In the low part of the near ground, the absorption coefficient of the infrasound signal of the debris flow is roughly $\alpha = 10^{-4}$dB/km. Infrasound waves can transmit at a speed of 344 m/s in air at room temperature with low frequency of infrasonic wave, small absorption and long propagation distance of the wave. In the medium, the propagation can take advantage of the unimpeded transmission through tiny gaps. According to its typical characteristics in the process of debris flow movement, infrasound wave is determined as a monitoring content of the debris flow disaster early warning.
   (4) **Mud level of the gully**. Given that HSR generally crosses debris flow gullies with bridges and culverts, it is necessary to determine the peak flow under the corresponding frequency of debris flow and the size of the flow section of bridges and culverts according to the passing capacity of debris flow gullies. The mud level warning threshold of debris flow at different warning levels should be determined according to the traffic safety of high-speed railway while taking the flow, hydraulic radius, cross-section flow area and longitudinal slope of the gully bed into account at specific section.

II. **Early warning process of the HSR debris flow**. The debris flow disaster warning system for the HSR safety operation selects four parameters, the real-time rainfall, rainfall infiltration depth of the source soil, mud level of the gully

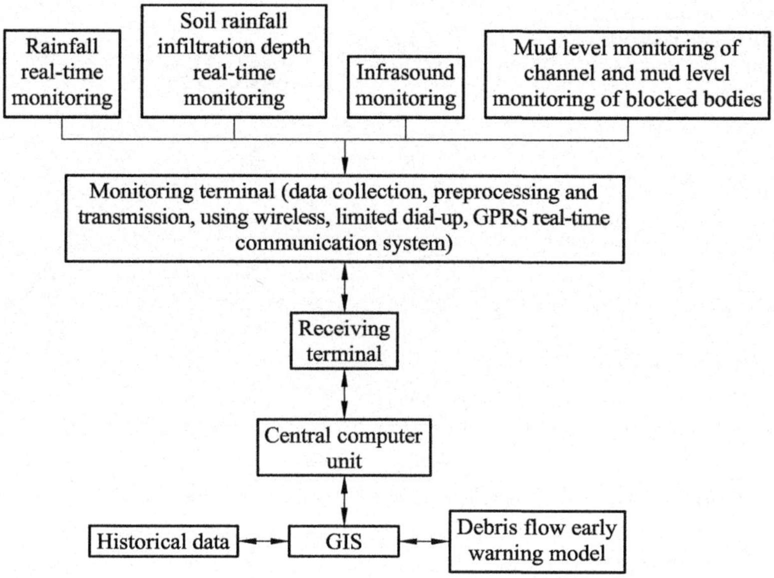

Fig. 6.8 Debris flow warning system along the HSR

and infrasound pressure value, as early warning indicators. By adopting systematic engineering methods and modern information technology means on the monitoring of the real-time rainfall information, rainfall infiltration information of the source soil, debris flow infrasound information and mud level information, the monitoring and transmission technology of the debris flow disaster along the HSR in the mountainous area that adapts to the harsh environment are thus developed. Now with a completed network system of debris flow monitoring and early warning system along mountainous railway lines based on multi-element and multi-channel data network and information database, the debris flow monitoring information can be mutually verified, and the early warning of debris flow can improve the accuracy of debris flow early warning and reduce the probability of false alarm and omission based on stable information sources. The structure of the early warning system for the HSR debris flow is shown in Fig. 6.8.

6.4.2 Data-Aware Model for the Geohazard Warning

As the most widely used Artificial Neural Network (ANN) model, the Backpropagation (BP) neural network is a nonlinear information processing system consisting of a large number of widely interconnected neurons. It can imitate the structure and function of the human brain. Therefore, this chapter uses BP neural network to build a geohazard warning model for the HSR safety operation.

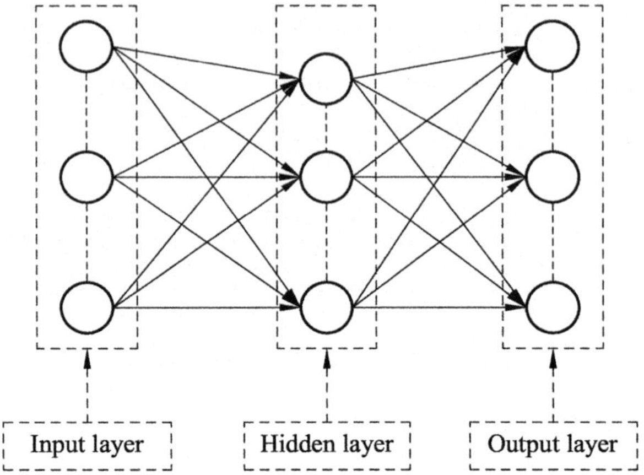

Fig. 6.9 Structure diagram of the BP neural network algorithm

(1) **Principle of the BP neural network**. BP neural network is a multi-layer feed-forward neural network based on BP error backpropagation algorithm. The neurons in the network are arranged hierarchically into three layer, the input layer, hidden layer and output layer according to their functions. The input layer and output layer have only one layer, while the hidden layer can have one or more layers. Its structure is shown in Fig. 6.9. Each neuron feeds forward only to all neurons in the next layer and is completely connected in each layer, while the neurons in each layer are not connected to each other. BP neural network adopts sigmoid-type transfer function, in two common forms.

$$f(x) = \frac{1}{1 + e^{-x}}, \tag{6.5}$$

$$f(x) = \frac{e^x - e^{-x}}{e^x + e^{-x}}. \tag{6.6}$$

The sigmoid transfer function has the function of nonlinear amplification coefficient, which can turn an input signal from negative infinity to positive infinity into an output between − 1 and 1. The amplification factor is smaller in the larger input signal in the HSR system and is larger in smaller input signals. The training of BP neural network is divided into two processes: forward propagation of information and backward propagation of error, as shown in Fig. 6.10. Therefore, the BP neural network is used to construct the data aware model of the HSR geohazard.

(2) **BP neural network algorithm**. According to the forward propagation principle of BP neural network, the data aware model of the HSR geohazard is:

Step 1: The input variables for the data aware of the HSR geohazard are set as:

$$X_k = (x_1, x_2, \ldots, x_n), \tag{6.7}$$

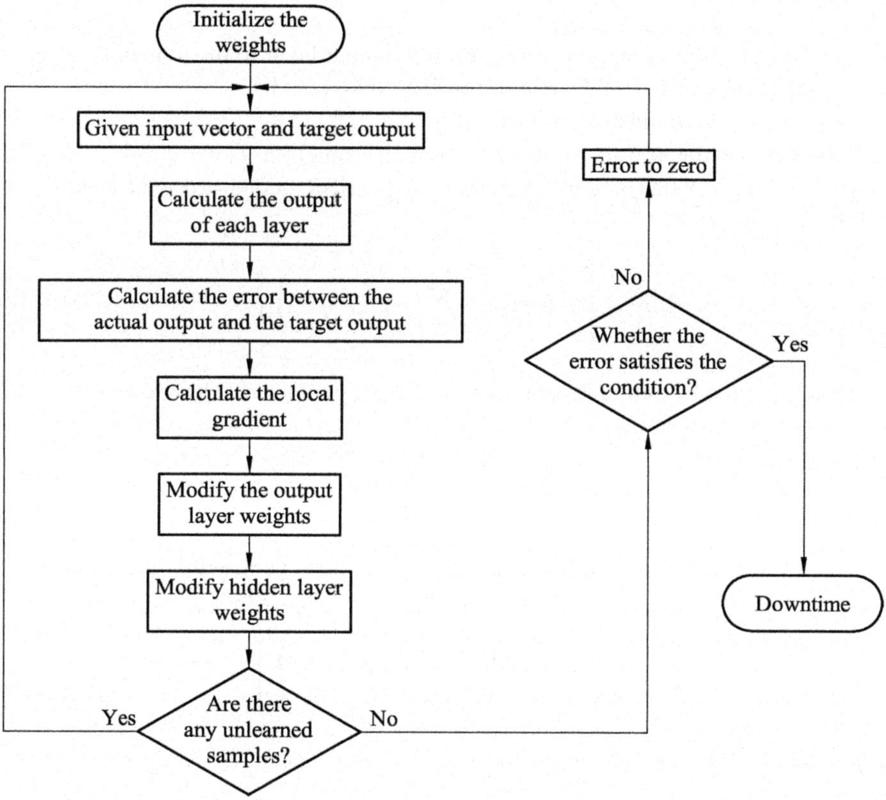

Fig. 6.10 Block diagram of BP neural network algorithm

where

$k = 1, 2, \ldots, m$, and m is the number of training samples;

n is the number of input layer cells;

Step2: The output vector of the corresponding input pattern of the HSR geohazard data aware is:

$$Y_k = (y_1, y_2, \ldots, y_q), \tag{6.8}$$

where

q is the number of output layer cells;

Step3: Inputs of each cell of the hidden layer for the HSR geohazard data aware are:

$$S_j = \sum_{i=1}^{n} w_{ij} x_j - \theta_j, \tag{6.9}$$

where $j = 1, 2, \ldots, p$;
w_{ij} is the connection weight from the input layer to the hidden layer;
θ_j is the threshold value of the hidden layer cells;
p is the number of hidden layer cells.

The sigmoid function is used for the transfer function $f(x) = \frac{1}{1+e^{-x}}$, then the output function of the hidden layer unit of the HSR geohazard data aware is as follows:

$$b_j = \left[1 + \exp\left(-\sum_{i=1}^{n} w_{ij}x_j - \theta_j\right)\right]^{-1}. \tag{6.10}$$

Input function of output layer unit for the HSR geohazard data aware is:

$$L_t = \sum_{j=1}^{n} w_{jt}b_j - \gamma_t. \tag{6.11}$$

Output function of the output layer unit for the HSR geohazard data aware is:

$$c_t = \left[1 + \exp\left(-\sum_{j=1}^{n} w_{jt}b_j - \gamma_t\right)\right]^{-1}, \tag{6.12}$$

where $t = 1, 2, \ldots, q$;

w_{jt} is the connection weight from the hidden layer to the output layer;
γ_t is the unit threshold of the output layer, so far, a forward transmission process is completed.

6.4.3 HSR Geohazard Warning System

The current monitoring of the HSR geohazards refers to setting up monitoring points along the HSR, installing rain gauges, infrasonic monitors or ultrasonic mud level meters and collection units and other units to collect the geohazard data in real time. The early warning system will issue an alarm when the relevant data exceed the alarm value on the basis of systematic analysis of the collected data. Timely measures shall be taken after the user confirms the alarm information and site conditions.

(1) Early warning system for the HSR debris flow disaster monitoring. In the weather with frequent geological disasters, the operation of high-speed trains is often delayed out of HSR safety, resulting in the instability of the HSR operation. Now, through the comparison of domestic and foreign railway geological monitoring and early warning systems, a HSR geological disaster monitoring and prediction system is designed. Components are as follows:

① The design of the debris flow information collection interface around the HSR based on rain gauge, infrasonic monitor or ultrasonic mud level meter, rainfall infiltration depth meter and mud level meter;
② The real-time transmission of sensor data based on 802.11-WIFI wireless transmitter module;

The upper computer data fusion processing center based on Visual Studio platform;

A safety assessment algorithm for the HSR operation based on the debris flow disaster with considerations of traditional debris flow disaster early warning and discrimination, considering the impact of debris flow disaster on train operation;

③ The information prompting function of the HSR operation safety assessment based on the mobile communication base station.

(2) **Types of the HSR debris flow disaster warning**. According to the characteristics of the debris flow disaster and the HSR's own disaster resistance, the HSR safety operation requirements, the early warning of debris flow disasters of different levels along the HSR line in mountainous areas can be divided into seven warning types (as shown in Table 6.6): no warning, suggestive warning, primary warning, formative warning, non-catastrophic warning, catastrophic warning and catastrophic advanced warning.

6.5 Summary of This Chapter

This chapter analyzed the mechanism of geological disaster impact on the HSR and the mode of disaster formation on the basis of summarizing the early warning system of the HSR debris flow disaster at home and abroad. Combined with the formation, movement and disaster mechanism of the debris flow disaster along the HSR, four parameters, real-time rainfall in the debris flow basin, rainfall infiltration depth of soil in the material source area, infrasound information and mud level in the gully, are proposed as mountain debris flow disaster early warning parameters in building the disaster early warning system.

The HSR geological disaster early warning system classifies the debris flow disaster early warning along the HSR into the following types: suggestive early warning, formative early warning, non-disaster early warning and disaster early warning, and proposes rainfall in the corresponding watershed, rainfall infiltration depth of the source land, debris flow infrasound and mud level of the debris flow gully as the threshold values of early warning indexes for key parameters. There is time left for dispatchers to make decisions and determine reasonable and appropriate countermeasures as warns will be given before the debris flow disaster affects the normal traffic of the line. Therefore, the system will play a very good role in preventing disasters and ensuring the safety and efficiency of the HSR traffic.

Table 6.6 Warning level of the debris flow disaster

Early warning level		Define the scope	Early warning measures
Level 1	No early warning	Railway bridges and culverts are in safe working condition	No warning issued and the train passed through the area at normal speed
Level 2	Suggestive warning	A possibility of disaster according to the detection information even though railway bridges and culverts are in a safe working state	Trains are running normally and the system is required to report debris flow hazards every 45 min
Level 3	Primary early warning	Small possibility of disaster while the railway bridges and culverts are in safe working condition	The railway bureau, works section, works area or their authorized units will issue the primary warning of debris flow to the high-speed trains passing the dangerous area, and the system shall make a debris flow disaster report every 30 min while the trains are running normally
Level 4	Formative early warning	High possibility of disaster while railway bridges, culverts and channels are everywhere in the critical working state of disaster, debris flow is in the formation state, but it has not reached the state of disaster	The railway bureau, works' section, works' area or its authorized units will issue a debris flow formation warning to the high-speed trains passing the dangerous area, and the trains will run at a reduced speed of 170 km/h
Level 5	Non disaster warning	While railway bridges, culverts and channels are in the dangerous working state of disaster and debris flow has been formed, but its scale is not enough to cause disaster	The railway bureau, works section, work area or its authorized units will issue a non-disaster warning of debris flow to the high-speed trains passing the dangerous area, and the trains will run at a reduced speed of 120 km/h and observe the passing area slowly
Level 6	Disaster warning	High possibility of disaster while railway bridges, culverts and channels are in a dangerous state of overflow	The railway bureau, works' section, work area or its authorized units will issue a disaster warning of debris flow to the high-speed trains passing the dangerous area, and the trains will run at a reduced speed of 80 km/h and observe the passing area slowly
Level 7	High disaster warning	A disaster has formed and railway bridges, culverts and channels are in extremely dangerous over-current state	The railway bureau, works' section, work area or its authorized units will issue a high warning of debris flow disaster to the high-speed trains passing through the dangerous area and perform interval locking and stop the train

Bibliography

1. Bian Y (2015) Research on wind and rain alarm and optimal disposal of high-speed rail disaster prevention system. Technol Innov Appl 10(8):46–47
2. Chen Z (2009) Formation and prevention measures of debris flow on the Nanning-Kunming Railway. Railway Oper Technol 15(4):26–30
3. Li Q (2011) Landslide risk assessment for planning railway network in Guizhou Province. J Central South Univ 42(10):3170–3175
4. Li C, Wang L, Liao K (2014) Study of early warning mechanism of debris flow along railway line in mountainous areas. Chin J Rock Mech Eng 33(2):3810–3816
5. Peixer MA et al (2021) Running safety evaluation of a train moving over a high-speed railway viaduct under different track conditions. Eng Fail Anal 121:105133
6. Bo T (2013) Early warning scheme of debris flow along high-speed railway. Chin Railways 51(10):11–16
7. Bitigova D, Bekzhanov D, Bekzhanova S (2021) Ensuring train safety on high-speed rail-highway. ВестникКазАТК 118(3):28–33
8. Li C, Hu X, Wang L (2011) Automatic monitoring and early warning system of mud-position of debris flow along railway line in mountainous area. J Nat Disast 20(5):74–81
9. Li C, Hu X, Wang L (2012) Design of a rainfall infiltration depth detector and its application in debris flow source area. Bull Soil Water Conserv 32(2):168–171

Chapter 7
A Crosswind Disaster Warning System for the HSR Safety Operation

Buoyancy and pitch moment generated during the running, of high-speed trains increases with the running speed, making the train in a "floating" state. If the high-speed train is also subject to strong crosswind, the possibility of train derailment, overturning and casualties will increase (Fig. 7.1). The HSR crosswind monitoring system has become an important guarantee to ensure the HSR safety operation and reduce or eliminate the railway traffic accidents caused by crosswind for the HSR safety operation. The disaster prevention and safety monitoring system (including crosswind monitoring system) on the newly built HSR in China vary according to the natural environment and geographical conditions of the line to ensure the HSR safety operation. For example, the disaster prevention and safety monitoring system established by Beijing–Shanghai HSR includes subsystems such as the wind monitoring, rainfall monitoring, earthquake monitoring and foreign object intrusion monitoring.

The crosswind, especially the lateral wind, has a greater impact on running high-speed trains, which increases with the train speed). Lateral wind will affect the safety, stability and comfort of train operation. The strong lateral winds will produce the lateral aerodynamic force that may cause the train, especially on extra-large bridges, high embankments or windy lines, to swing across the limit, fall off the track, and even overturn and casualty accidents. Therefore, the establishment of the HSR crosswind monitoring and early warning system is of great practical significance to prevent crosswind accidents and ensure the HSR safety operation.

7.1 Studies on the Crosswind Warning System

Trains are often delayed due to safety concerns in windy days, affecting passenger travel. The traditional crosswind control measures are based on the experience of handling accidents in the past and summary analysis. However, the traditional

Q. Hu, *Natural Disaster Warning System for High-Speed Railway Safety Operation*, Advances in High-speed Rail Technology, https://doi.org/10.1007/978-981-99-7115-2_7

Fig. 7.1 Strong winds cause "5807 train accidents in Xinjiang"

measures, in spite of comprehensive consideration of wind direction and wind speed, are poorly applicable as they only work in areas where the natural wind blows in a specific direction with relatively small change of wind direction. Therefore, by studying railway crosswind monitoring system at home and abroad in this chapter, the system of HSR crosswind monitoring and early warning suitable for China's national conditions is constructed based on GIS and sensor technology.

7.1.1 Crosswind Monitoring and Warning System in Germany

In the 1990s, in order to improve traffic safety, German railway companies increased their investment in railway risk reporting equipment and installed the risk detection equipment on high-speed lines, new lines and existing lines. These devices are used to monitor, identify and report irregularities in rolling stock, and also identify and timely report the extent to which the environment along the line affects trains. Figure 7.2 shows the process flow of crosswind monitoring in Germany.

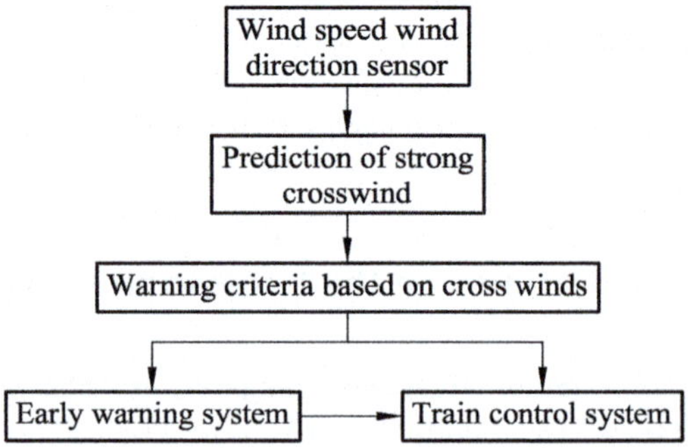

Fig. 7.2 Early warning framework for strong wind monitoring in Germany

(1) **Now casting system for the HSR**. One aspect of the development of the modern high-speed train is to reduce the axle weight. However, due to the reduction of axle load, the train is more vulnerable to the impact of crosswind when running at high speed. Therefore, in 1998, Deutsche Bahn developed an intelligent short-term wind alarm system "Now casting" whose main function is for the train to run at the maximum allowable speed of trains affected by the crosswind through the prediction of strong crosswind on the line.

For the HSR running under strong crosswind, the higher the speed of the train means the stronger the sensitivity to the crosswind, and the more unsafe the train will be. Therefore, the speed deceleration is used to ensure the safety of HSR. This principle is carried out in the Now casting system in which trains will be forced to slow down when the wind speed exceeds a certain value. Figure 7.3 shows the working block diagram of the German HSR Now casting system. When the train is running, the wind measurement points along the line predict the wind direction and wind speed in front of the train through the prediction model and determine the maximum running speed of the train according to the prediction, train type and line condition while continuously transmitting the latest wind signal to the system, leaving the speed under system control.

The German Now casting system can cover all areas where the critical wind speed exceeds 20 m/s and gives an alert 120 s in advance. Once the alarm is issued, the affected trains must reduce their speed to the speed required by the alarm within this period of time to ensure safety. Currently, the communication range of a single Now casting subsystem with each train control system is about 50–100 km.

(2) **New MAS90 system for the HSR.** In the 1990s, Deutsche Bahn's danger detection equipment was networked with the MAS90 reporting system, which is incorporated into the regional traffic control center that is currently under heavy

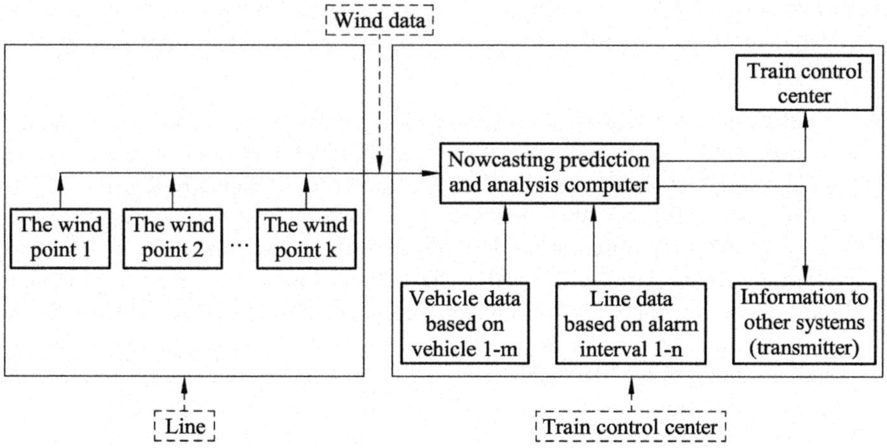

Fig. 7.3 German "Now casting" system

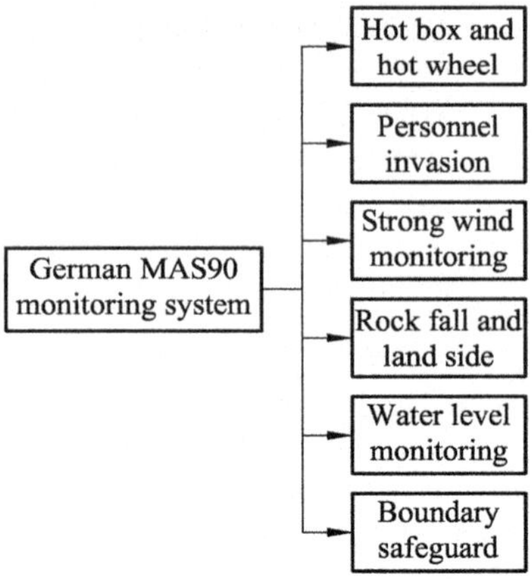

Fig. 7.4 Germany "MAS90" environmental warning system

construction by Deutsche Bahn, to diagnose and count the supervised objects and equipment (Fig. 7.4).

7.1.2 Crosswind Monitoring and Warning System in Japan

Japan attaches great importance to the development of high-speed trains and has built a nationwide railway network at a very early stage. Japanese scholars have made a lot of achievements through comprehensive research on the safe operation situation of the HSR, especially on the characteristics of wind, such as the relationship between the change of instantaneous wind speed and the control wind speed. If taking the factor of wind speed increase into account, it is believed that the control wind speed should be less than the critical overturning wind speed, and the difference between the two is determined by two factors: the distance between the vehicle location and the wind measurement point and the time required for the train to pass the location completely.

(1) **Traditional crosswind warning system in Japan**. In the early days, the danger of crosswinds to Japanese railways was reduced by wind control measures. The concise rules of wind control currently applied in the Japan Railway-East (JR-EAST) are as follows: the control starts immediately at the moment when the wind speed at the observation point exceeds the set standard; the control ends 30 min after the last wind speed at the observation point exceeds the set standard. Human resources are also deployed in part of the traditional strong wind early warning system of Japanese railways. With the development of science and technology, the JR-EAST was the first to obtain and apply the required disaster information from the computer system of the Japan Meteorological Agency and the PreDAS (Disaster Alert System) introduced in 1990 to railway operation. This greatly improved the warning speed of the warning system and reduced the errors that could be caused by human factors. Subsequently, the wind

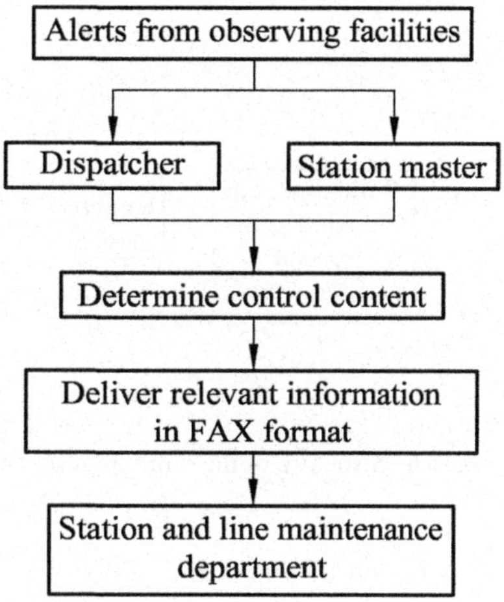

Fig. 7.5 Flowchart of Japanese early disaster prevention system

warning system was gradually developed and integrated into the transportation management system (Fig. 7.5).

In light of the overturning accident of the Uetsu Line on December 25, 2005, the JR-EAST imposed stricter speed limits on all speed-restricted sections of its lines in the crosswind environment on January 19, 2006: The train speed is limited to 25 km/h at a wind speed of 20 m/s and stops at a wind speed of 25 m/s. Since the more frequent speed limit operation due to strong wind has made it very inconvenient for passengers to travel, the JR-EAST has taken two measures to reduce the impact of speed restrictions on rail transportation in the section of the line near Tokyo by setting up the windbreak fence, and installing a crosswind warning system in the section of crosswind, the former of which can the frequency of restricted operation, while the latter of which can shorten the time of restricted operation.

(2) **WINDAS system for the HSR.** The technical system used in the Japanese WINDAS (wind profiler) system is based on the characteristics of the Japanese weather system of the central and western regions of Japan given the frequent occurrence of heavy rain here. The average station interval is 130 km. The wind profiler equipment, manufactured by Mitsubishi Japan, with high transmitting power, high antenna gain, advanced data pulse compression technology, clutter protection, antenna cover equipment, automatic data quality control, etc., is operated remotely by the Tokyo control center. The computer in the center calculates the u, v and w components of the wind after. Average Doppler velocity vertical profile and the signal-to-noise ratio data of five beams every 10 min are all sent to the control center. In general, Japan's wind profiler technical device is mainly designed to provide initial wind field data for the mesoscale numerical prediction model. The WINDAS system can help solve the problems in the traditional wind control measures from two perspectives. One is that the early start of wind control makes the train decelerate to a safe running speed within the predicted time, and the prediction of wind speed by the WINDAS system can

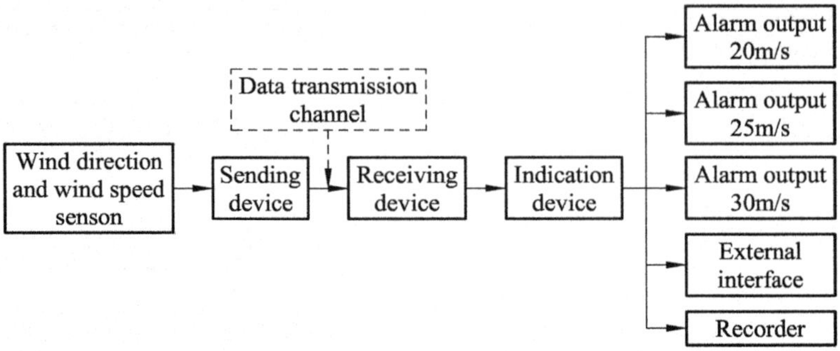

Fig. 7.6 Structure of the wind direction and wind speed subsystem of Japanese HSR

be ten minutes to dozens of minutes in advance; the other is that the predicted wind speed makes it possible to end the wind control within a definite time, without requiring strict regulations for 30 min. However, the WINDAS system still misses a fixed wind measuring point near the line and is very expensive in costs of investment, operation and maintenance.

(3) **Shinkansen crosswind warning system.** Japan's Shinkansen uses a crosswind warning system, as shown in Fig. 7.6 with a certain wind direction and wind speed monitoring subsystem. The wind direction and anemometer convert analog electrical signals into digital signals through its attached converter and send them to the analysis, recording and display device through a pair of cables through their respective signal transmission devices. When the wind speed reaches a certain value, the central control center is automatically notified to control the train to slow down or stop based on the line conditions, wind resistance of the train and the surrounding environment.

Shinkansen has also established control rules for the crosswind operation mainly based on instantaneous wind values. Most of researches in Japan on control release wind speed still use some empirical formulas in spite of large research cases. For example, the empirical formula: the wind speed at which the operation control is released = instantaneous wind speed instantaneous wind rate (where instantaneous wind rate = maximum wind speed/average wind speed), or the specified time to end the wind control is 30 min later.

7.1.3 Crosswind Monitoring and Warning System in China

Given its vast territory, complex terrain and frequent natural disasters the most serious natural disaster for HSR is the crosswind disaster, especially strong crosswinds in lines like Lanzhou-Xinjiang Line in China. High-speed trains traveling under crosswind conditions are not only affected by the aerodynamic resistance in the direction of running but also by the angle between the direction of the line and the main wind direction of crosswinds. In special places (such as extra-large bridges, high embankments, viaducts, passes, canyon areas, etc.), the wind will have a narrow pipe effect or a growth effect. When the included angle between the line direction in special environment and the main wind direction of strong wind is 75–95°, the aerodynamic force of strong crosswind on the train will be greatly increased, leading vibration and even overturn in serious cases of a train. Therefore, it is very important to build a crosswind detection system for high-speed railway in China.

7.1.3.1 Crosswind Warning System for the HSR Safety Operation

With consideration of the impact of natural disasters on high-speed train travel safety, high-speed train safety operation control, and the principle of deployment to achieve the monitoring of the wind direction, wind speed, rainfall, air pressure, temperature and relative humidity, China's crosswind early warning system for the HSR safety operation can carry out comprehensive monitoring and early warning, realizing automatic control or manual control of car traffic. The basic process is shown in Fig. 7.7.

Under different types of crosswind conditions, the warning system can provide more reasonable information on the speed restriction instruction for the operation command and control system by monitoring the six elements, wind speed, direction, temperature, relative humidity, air pressure and rainfall of the crosswind, in real time, and importing the basic parameters and the judgment of the historical database, the database of wind monitoring points along the line as well as the embedded program, respectively.

In order to ensure the HSR safety operation in China under crosswind conditions, it is necessary to implement speed restriction and temporary stopping measures and develop short-time wind speed prediction models for different types of conditions on crosswind sections in the crosswind warning systems for the HSR safety operation and control for safe and efficient operation. China's HSR crosswind monitoring and early warning system are designed as a four-layer structure after comparing

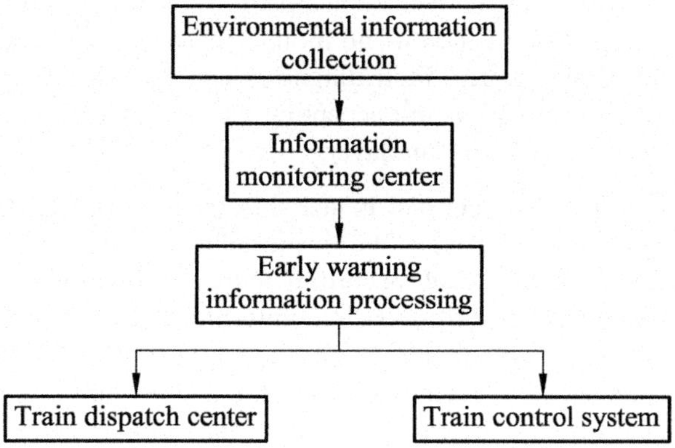

Fig. 7.7 Flowchart of the crosswind warning for the HSR in China

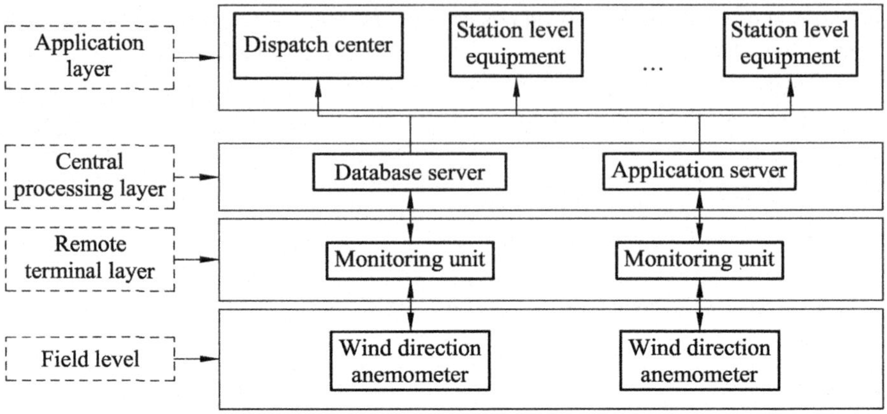

Fig. 7.8 Crosswind monitoring and warning system for the HSR in China

with foreign railway crosswind monitoring and early warning system. Its building structure is shown in Fig. 7.8.

Layer 1: The on-site layer of the HSR crosswind warning system. The sensor layer is installed with the aerovane to directly obtain the wind information along the HSR track.

Layer 2: The remote terminal layer of the HSR crosswind warning system. The monitoring unit equipment is installed in the GSM-R base station near the on-site detection equipment and placed in parallel with other communication cabinets. The monitoring unit equipment includes system host (microcontroller), UPS (uninterruptible power supply equipment), data receiving and sending module, relay combination module, lightning protection unit, network interface and cabinet and other equipment.

The main functions of the monitoring unit are as follows: uploading data to the central processing layer through the transmission network after collecting from the on-site monitoring equipment in real time, processing and storing it for a short period of time by the embedded acquisition software; monitoring and managing the status information of the on-site monitoring equipment while conducting self-tests, realizing fault alarm, fault diagnosis and fault location, uploading fault records and other information and accepting the centralized monitoring and management from the central processing layer.

Layer 3: The central processing layer of the HSR crosswind warning system. This layer is responsible for providing disaster prevention alarms and warnings of corresponding levels according to the information content after receiving f information transmitted from each monitoring, r, storing, analyzing and processing the data within the jurisdiction in real time. It also provides the information on speed limits, stoppages and other plans according to the rules of operation control and uploads the alarm information to the users of the application layer. This layer mainly consists of database server, a dual-computer hot standby mode with relevant monitoring data tables to store various system information, including wind direction and speed information, alarm information, equipment failure information and other historical records for inquiry, and application server responsible for collecting the status information from the monitoring unit host to the monitoring point and equipment status information, and storing various system information into the database server.

Layer 4: The application layer of the HSR crosswind warning system. It provides data display, printing and other functions to assist users to make scientific disaster prevention decisions, and interfaces to the train control system to ensure timely speed limit or stop of trains traveling in the wind zone.

At present, the monitoring of the crosswind is mainly to set up at monitoring points along the HSR, install wind speed and direction sensors and collection units, collect wind speed and direction data in real time and issue alarms when the data exceed the alarm value. After confirming the alarm information and site conditions, the user, generally railway dispatchers, can take timely countermeasures, such as slowing down or stopping the train or taking shelter. Leaving time for dispatchers to determine reasonable and appropriate countermeasures against strong winds, especially those warned in advance will play a very good role in preventing disasters and ensuring the safety and efficiency of traffic.

7.1.3.2 Wind Speed Limit and Speed Limit Standard

Information technology should be carried out in online monitoring monitor over the possible hazards and traffic safety disasters in real time on the basis of the preset threshold classification, or in issuing an overrun warning, and setting the standard for high-speed train safety operation crosswind warning signal in China according to the overall technical scheme of the high-speed railway disaster prevention safety monitoring system. As for the standards for wind speed limit and speed limit value, the provisions of TJS [2009] No. 212 document issued by the Ministry of Transport of China are shown in Table 7.1.

Compared to the wind speed limit standard of Guangzhou Railway Group released in 2012, it is found that the crosswind speed limit standard released by the railway

Table 7.1 Standard of the speed limit of the ministry of transport

Early warning level	Ambient wind speed	Speed limit standard
Level 1	$\leq$ 15 m/s	Normal speed operation
Level 2	15 m/s < Wind speed $\leq$ 20 m/s	$\leq$ 300 km/h
Level 3	20 m/s < Wind speed $\leq$ 25 m/s	$\leq$ 200 km/h
Level 4	25 m/s < Wind speed $\leq$ 30 m/s	$\leq$ 120 km/h
Level 5	> 30 m/s	Bullet trains are strictly prohibited from entering the wind zone

Table 7.2 Regulation No.157 of Guangzhou Railway engineering and electric [2012]

Early warning level	Ambient wind speed	Speed limit standard
Level 1	≤ 13.9 m/s	Normal speed operation
Level 2	13.9 m/s < Wind speed ≤ 17.2 m/s	≤ 250 km/h
Level 3	17.2 m/s < Wind speed ≤ 20.8 m/s	≤ 160 km/h
Level 4	20.8 m/s < Wind speed ≤ 24.5 m/s	≤ 120 km/h
Level 5	> 24.5 m/s	Bullet trains are strictly prohibited from entering the wind zone

department is earlier than the one released by Guangzhou Railway Group; the speed limit is based on the wind speed of 5 m/s rather than the actual value simulation results as the interval unit; and the speed control interval is up to 300 km/h instead of 250 km/h of Guangzhou Railway Group. It can be told that compared with the speed limit standard issued by the Ministry of Transport, the speed limit standard of the Guangzhou Railway Group is stricter as shown in Table 7.2.

7.1.4 Comparative Analysis of Domestic and Foreign Crosswind Monitoring Systems

The vehicle-mounted crosswind warning and control system in China, the Now casting system in Germany and the WINDAS system in Japan are all designed for safe operation of high-speed trains in crosswind conditions. In the case of strong crosswind, high-speed train operation under the control of these three systems can ensure better economy than traditional operation does. But the differences are as follows.

(1) **Wind measurement methods on the HSR lines are different in space**. Germany's Now casting system sets up a number of fixed wind measuring points along the line to predict the wind situation of the whole line by concentrating the wind signals together through certain communication means. In China, the vehicle-mounted crosswind early warning and control system sets the measuring points on the train to send the signal directly to the system. While the fixed wind measuring point can measure the wind speed and direction at this point very accurately, the multiple points can measure at the same time, establishing the complete wind environment monitoring on high-speed railway lines and providing a good basis for the prediction of crosswind. The prediction at the wind measuring points is more accurate as it only involves a function of time while the prediction between the wind measuring points is about function of time and space. The prediction ability of the system for dangerous points is also improved as the wind measuring points are often set up in places where the wind speed is easy to exceed the standard. However, fixed wind measurement points often require large investment and high operation and maintenance costs. The

installation of wind measuring points on the train, although with small investment, flexible application and convenient use, is good at accurately measuring the natural wind signal and eliminate the impact of train driving wind.

(2) **In terms of time, the prediction time is different**. Germany's Now casting system sends out a wind alarm 120 s in advance that lasts for at least 5 min. The WINDAS system in Japan can predict the wind speed from ten minutes to dozens of minutes in advance. The time for wind prediction of vehicle-mounted crosswind warning and control system in China has not been determined as it is still under development. The prediction time of wind speed and direction is limited by at least two aspects. On the one hand, the longer the prediction time is, the farther the next operation point, namely the prediction point, will be, and the lower reliability of the prediction. On the other hand, the shorter the prediction time is, the more likely the train will not decelerate to a safe speed within the prediction time, and thus the lower the safety. Therefore, the prediction time should be a compromise between the two, that is to increase the prediction time as much as possible on the premise of ensuring a certain degree of reliability.

7.2 Influence Mechanisms of the Crosswind on the HSR Safety Operation

Drivers should pay more attention to sudden crosswinds in some windy or wide areas, avoiding direction shift of trains l. A crash often happens when a high-speed train is subjected to crosswinds as the negative impact strengthens due to the high gravity of the overall center and the lateral area of such vehicles. The effect of crosswinds is intensified with the increase of vehicle speed, which explains why high-speed trains are often suddenly hit by strong crosswinds the moment they exit a tunnel, or when they drive toward a bridge or high embankment where the wind is running through.

7.2.1 Characterization of Crosswind Hazards

In the HSR system, trains in operation are mainly subject to locomotive traction, air resistance in the opposite direction of operation, air force from the side, subgrade force on the train, etc. These forces, together with the inertial force during accelerated operation, can keep the train in balance. But can also make the train off balance or even overturn once the force conditions or the operation state changes. For example, a high-speed train will lose its balance and may overturn in crosswind.

(1) **Reasons that the crosswind causes the HSR accidents**. There are two main reasons for accidents caused by strong winds during high-speed railway operation: The first is over high wind speed or large aerodynamic overturning torque acting on the train at even relatively low speed (e.g., around 100 km/h); the second is high travel speed of the train (e.g., above 200 km/h) against moderate

wind speed. When the wind acts on the train, the train wheel and track dynamics characteristics are easy to change, the consequentially large the derailment coefficient and load reduction rate will create train derailment and even overturning. With the increase of train speed, the aerodynamic of train has become a more prominent and important problem.

(2) **Rollover mechanism of the high-speed train caused by the crosswind**. e The buoyancy and pitch moment generated when the train is running at high speed increases along with the operating speed, making the train in a "floating" state, so is the effect of lateral wind on the train be. A high-speed train operating in a crosswind environment is subject to lateral force, lift force and drag force, so do overturning moment, headshaking moment and nodding moment caused by these three forces. Theoretically, a rollover accident occurs when the overturning generated by the lateral force on the train exceeds the train's heavy moment. As the outflow field around the train in the crosswind environment is different from the one without crosswind, it directly leads to a significant increase in the overturning moment of the train and the chance of train overturning accident.

(3) **Discomfort that crosswinds bring to high-speed trains**. Lateral wind will affect the safety, stability and comfort of train operation. Firstly, it will make the train resonate, causing discomfort to passengers and fatigue damage to the vehicle structure when reaching a certain speed; secondly, a strong lateral wind may cause the train to swing over the limit, fall off the track, and even overturn and casualty accidents due to increased lateral aerodynamic force to which the train is subjected; finally, the train flow field is obviously changed in some special crosswind environment, such as extra-large bridges, viaducts, embankments, hilly and mountainous areas, and curved lines where lateral aerodynamic force and centrifugal force are superimposed, resulting in significant change of aerodynamic force, and much higher possibility of train falling off track and overturning.

7.2.2 Key Parameters for Crosswind Monitoring

In the crosswind environment, the synthetic wind always acts on the vehicle with a certain side deflection angle. Due to the asymmetry of the flow field on the windward and leeward sides of the train, the vehicle is bound to undergo the action of lateral aerodynamic force and aerodynamic moment. In the vertical direction, due to the flow field distribution of the streamlined front, middle body part and streamlined rear part of the high-speed train is subjected to the action of larger nodal moments, especially for the head and tail cars. Due to the difference in the negative pressure on the top and bottom surfaces of the train, the train is subjected to vertical lift. Whose upward or downward direction is decided by the cumulative result of air pressure on the surface of the train. Therefore, a perfect and reliable wind speed monitoring and alarm system is established in the dangerous sections of train operation, such as extra-large bridges or viaducts, embankments, high embankments and windy lines, to ensure the safety of high-speed trains in the corresponding environment.

(1) **Side force parameters of the high-speed train**. The lateral force in the HSR safety operation in crosswind refers to the resultant force formed along the transverse direction by the distributed pressure and shear stress on the vehicle surface when the high-speed train is running, or say, the sum of the transverse force of air pressure difference on the vehicle surface and the transverse force of friction. The functional relationship is as follows:

$$F_{\mathrm{S}} = \frac{1}{2}\rho C_{\mathrm{S}} A_{\mathrm{S}} U^2, \tag{7.1}$$

where

- ρ Is the air density in the HSR safety operation;
- C_{s} Is the side force coefficient of the high-speed train in the HSR safety operation;
- A_{s} Is the side projection area of the high-speed train in the HSR safety operation;
- U Is the synthetic wind speed in the HSR safety operation.

It can be seen from (7.1) that the side force of the head car and the middle car peaks when the wind direction angle is 90 m/s in the HSR safety operation. When the wind direction angle is acute, the side force is greater than that of the obtuse angle about 90° of symmetry. The side force under each wind direction angle increases along with wind speed and vehicle speed, and the side force increases most obviously when the wind direction angle is 90 m/s.

(2) **Lift parameters of the high-speed train.** Lift1 refers to the resultant force formed along the vertical direction by the distributed pressure and shear stress on the vehicle surface when the high-speed train is running, that is, the sum of the air pressure difference lift and friction lift on the vehicle surface along the vertical direction in the HSR safety operation running in crosswinds. The functional relationship is as follows:

$$F_{\mathrm{L}} = \frac{1}{2}\rho C_{\mathrm{L}} A_{\mathrm{h}} U^2, \tag{7.2}$$

where

- C_{L} Is the side force coefficient of high-speed trains in the HSR safety operation;
- A_{h} Is the horizontal projection area of high-speed trains in the HSR safety operation.

(3) **Tipping moment parameters of the high-speed train**. In the HSR safety operation, high-speed vehicles will bear the overturning moment around x-axis under the action of lateral aerodynamic force and lift force. The function relationship is as follows:

$$M_x = \frac{1}{2}\rho C_{Mx} A_s H U^2, \tag{7.3}$$

where

H Is the height of the high-speed vehicle;
C_{Mx} Is the coefficient of overturning moment in the HSR safety operation.

It can be seen from (7.2) that in the HSR safety operation, the overturning moment for the head car and intermediate car is at maximum when the wind direction angle is 90 m/s. The overturning moment under each wind direction angle increases along with the wind speed and vehicle speed and peaks when the wind direction angle is 90 m/s.

(4) **Shaking moment parameters of the high-speed train**. In the HSR safety operation, high-speed vehicles will bear the shaking moment around the y-axis under the combined action of aerodynamic resistance and lift. The functional relationship is as follows:

$$M_y = \frac{1}{2}\rho C_{My} A_h H U^2, \tag{7.4}$$

where

C_{My} Is the shaking moment coefficient in the HSR safety operation.

It can be seen from (7.4) that in the HSR safety operation, of the shaking moments of three cars, the head car of high-speed train is the largest, followed by the tail car, and the middle car is the smallest.

(5) **Nodding moment parameters of the high-speed train**. In the HSR safety operation, high-speed vehicles will bear the nodding moment around the z-axis under the combined action of aerodynamic resistance and lateral aerodynamic force. The function relationship is as follows:

$$M_z = \frac{1}{2}\rho C_{Mz} A_h H U^2, \tag{7.5}$$

where

C_{Ms} is the nodding moment coefficient in the HSR safety operation.

(6) **Selection of the wind speed and direction indicator**. In the HSR safety operation, two wind speeds are used in the strong wind warning: One is the prediction of the maximum wind speed of the operation section that the EMU will enter in advance for a short time according to the continuously monitored wind speed so as to obtain the predicted maximum wind speed. The other is the instantaneous wind speed calculated by the meteorological model algorithm of the actual monitoring. But operation control shall be implemented no matter which wind speed value exceeds the warning threshold. Corresponding warning signals will be sent out when one of the measured instantaneous wind speed or the predicted wind speed reaches the different speed limits in this section; the early warning signal will be removed, and the high-speed train operation control curve will be drawn under strong wind conditions when both the measured instantaneous wind speed and the predicted wind speed are lower than the threshold value, t to achieve the goal of safe and efficient traffic.

According to the working principle, the aerovane can be divided into two categories, namely mechanical and ultrasonic, and two types of mechanical aerovanes: three-cup type and propeller type. However, it is difficult for aerovane to meet the high-reliability requirements of the HSR disaster prevention and monitoring system due to its rotating parts, wear loss, wind and sand loss, freezing, rain and snow interference, and heavy maintenance workload in spite of simple structure and low prices. The relatively expensive ultrasonic aerovane with high precision and reliability is recommended in the HSR disaster prevention and early warning system.

(7) **Setting principle of the monitoring point**. Crosswind hazard is closely related to regional conditions and geomorphology. However, since the national meteorological department can only provide a wide range of meteorological profile instead of forecasting the specific section, it is necessary for high-speed railway to set up wind monitoring points in sections that can reflect the characteristics of local strong winds, such as stations, extra-large bridges, overhead lines and substations in areas prone to strong winds or sudden strong winds, as well as open areas with long wind periods. Wind monitoring points shall be set in the wind outlet area with strong wind, and the data measured at the selected monitoring points shall represent the actual air volume in the area. Ultrasonic wind direction anemometer for monitoring vertical ground longitudinal wind speed, if necessary, can also be set on the bridge for the HSR safety operation.

7.3 Wind Speed Prediction Mechanism for the HSR Safety Operation

The change of wind speed is a rolling dice in the HSR safety operation. The HSR crosswind prediction refers to the use of instantaneous wind data that have been recorded in the current minutes or hours to predict the dangerous wind speed that will appear in the next 3 –10 min in a specific area, with which the train can slow down in advance. The safety of train operation will be improved if high-speed trains slow down to a safer speed (such as below 160 km/h) after receiving an alarm signal.

At present, high-speed railway departments in China, France, Japan, Germany, and other countries are working on the monitoring and forecast of the strong wind on high-speed railway. At present, there are seven prediction models, meteorological model, remote sensing prediction model, expert system prediction model, time series analytic prediction model, neural network prediction model, time series analytic prediction model combined with neural network prediction model, meteorological model combined with neural network prediction model, etc., as shown in Table 7.3.

Table 7.3 Wind speed prediction model

Number	Prediction model	Characteristic	Applicability of safe operation control
1	Meteorological prediction model	A method for predicting macroscopic detailed meteorological physical models considering the spatial distribution and topography and other factors at each location	It is suitable for sections with significant air pressure difference
2	Remote sensing prediction model	Predict the wind speed at various points using special observation devices that can determine the distribution of wind direction and wind speed on the surface	It is suitable for railways with linear scope and for large scale
3	Expert system prediction model	Make predictions based on historical data and information	It is necessary to leave the time to build a knowledge database based on predictions
4	Time series analytic prediction model	Look at the interdependence of data collected over time	It can predict objectively and quantitatively; unpredictable bursts of strong winds
5	GRNN neural networks predict patterns	Sampling wind speed data is used to modify the network directly, without recalculating parameters	It is suitable for making objective and quantitative prediction in passenger dedicated lines
6	Neural networks predict patterns	The neural network was selected to output the wind speed predicted in cold wave and gale weather conditions along the HSR 5 min in advance	It is suitable for cold wind with large influence range along the HSR
7	Meteorological model and GRNN neural network are combined to predict the model	Combined with k coefficient and wind speed data of each monitoring point, the meteorological model algorithm was used. Output the predicted wind speed under thunderstorm and gale weather conditions along the HSR 2 min in advance	It is suitable for short-time thunderstorms and crosswinds with complicated terrain. The longer the prediction period in advance, the greater the error will be

The crosswind warning system for the HSR safety operation controls trains works on the basis of the maximum predicted wind speed, whereby the system initiates restrictive operation measures when the predicted wind speed exceeds the control value. Compared with the control based on the measured wind speed, the safety of trains in the crosswind environment is more assured by train control under the predicted wind speed. Moreover, the crosswind warning system can lift the speed control when both the measured wind speed and the maximum predicted wind speed are lower than the control value, shortening the time by 20–30% on average compared with the crosswind control method and significantly improving the efficiency of the HSR operation.

In the HSR safety operation, the most direct way to predict wind is to use weather forecast simulation analysis. However, the meteorological model is very complex and the results are not accurate enough due to many factors such as special air flow on the ground and terrain around the railway line, making it fail to meet the real-time and reliable requirements of the HSR for the crosswind prediction. Different from ordinary weather forecast, the HSR crosswind forecast is a meteorological forecast within a specified small-scale range (time and interval), and a method of extrapolation forecast according to the law of wind while considering the correlation between wind and time and space scale and using wind data observed in the region. The key technology of the HSR crosswind prediction is to achieve the prediction of more than 3–5 min in advance by selecting appropriate methods to establish an extrapolation model for the measured wind speed along the line.

7.4 Crosswind Warning Thresholds for the HSR Safety Operation

Many countries around the world have conducted research on the crosswind for high-speed trains and have identified their own crosswind limitation standards for the HSR in the actual environment. For example, earlier studies in Europe on the crosswind safety issues have helped the region formulate the Technical Specification for Interoperability (TSI). In order to determine the relationship between crosswind speed and train speed limits, according to TSI, the wind speed characteristic curve refers to the wind speed characteristic curve or the reference wind speed characteristic curve.

7.4.1 Crosswind Operation Mechanism of German HSR

Since 1994, the Deutsche Bahn has been investigating the effects of crosswinds on these lightweight vehicles. The study found that there are 3 days a year when the maximum wind speed of strong wind exceeds 25 m/s every year on the embankment or bridge on average during which the continuous acceleration from the existing line running at 200 km/h will lead to the overturning of vehicles. Therefore, German researchers have listed four steps to ensure the safety of train operation after studying the strong wind warning system:

Step 1: Determine the maximum transverse wind speed of the vehicle. By applying a multi-body dynamics approach to analyze the wheel-rail contact forces in the case of strong crosswinds, the aerodynamic coefficients for measuring in wind tunnel tests, and taking influencing factors of the characteristic parameters of the drive, the geometric unevenness of the track, and the line superstructure into account, the characteristic wind curve is concluded.
Step 2: Probability assessment of the hazard for a specific line. The function of function determining the local allowable wind speed based on the characteristic wind curve and the possibility of a strong crosswind through statistics of wind speed and direction, route direction, track position (embankment, bridge), roughness of line surface and surrounding ground, and trackside buildings are included to determine the probability. At the end, the function can be expressed as: probability of the vehicle × probability of the strong wind.
Step 3: Risk assessment for the HSR safety operation, in which the final operating speed between two stations based on the function should be determined.
Step 4: Develop safe operation measures. If the characteristic wind curve exceeds the target value, measures, such as reducing the speed, installing wind-blocking walls, adjusting the structure of the car body (reducing the center of gravity height, improving the aerodynamic characteristics, increasing the counterweight), and rectifying the level of the line, must be taken, all of which, generally are optimized to achieve the target values more economically.

Germany has also developed a short-term crosswind forecasting model based on a number of anemometers installed along the line. The model can issue commands to regulate the operating speed of the vehicle after transmitting the vehicle type, line conditions and wind speed through the rail line to the control center. The system also gives the train enough time for speed limit operation as it can forecast the crosswind speed 2 min in advance. In the German HSR construction, some crosswind sections tend to design windproof walls but ignore the influence of wind direction.

Table 7.4 listed in the Hanover–Würzburg HSR crosswind operating limits, and Japan's speed limit standard comparison difference is significant. While it is true that German passenger trains are not restricted by the direction of the wind, they still install wind speed and wind direction sensors in specific locations and on long-span bridges, and they send the data to the disaster prevention alert system to regulate freight trains.

Table 7.4 German high-speed railway high wind operation limit standard

Wind speed v/(m/s)	$20 < v \le 25$	$25 < v \le 32$	$v > 32$
Passenger train	Continue to run	Continue to run	Continue to run
Freight train	The speed limit for trains prone to being blown over is 80 km	All trains have a speed limit of 80 km	Stop running

7.4.2 Crosswind Operation Mechanism of Japan's HSR

In the early days, Japan's railroads included wind control methods to mitigate crosswind damage. The following are the brief guidelines of risk control now implemented in JR-EAST: The concise rules of risk control currently applied in JR-EAST are as follows:

Warning rule I. The control starts immediately at the moment when the wind speed at the observation point exceeds the set standard.
Warning rule II. When the final wind speed at the observation location exceeds the established criteria, the control expires 30 min later.

JR-EAST provided new wind control procedures for trains that incorporate both wind direction and wind speed to prevent vehicle derailment in crosswinds. JR-EAST found that the current speed limit measures have some shortcomings. For example, after detecting a crosswind, the speed limit directive was sent for 30 min. The directive was nonetheless effective even if there was no crosswind. JR-EAST developed a crosswind warning system based on time series analysis and the Lab view to address this issue. Based on the time series analysis of the speed data collected by meteorological meters positioned along the tracks, this system can predict the wind speed in the next 30 min and issue speed limit orders. Japan provides the unique algorithms utilized in the computation in the meantime.

Step I. Make the maximum wind speed time series based on the observed continuous wind every 3 min.
Step II. Estimate the trend of the maximum velocity time series using Kalman filtering.
Step III. Calculate the point estimation of the maximum wind speed at each moment.
Step IV. Estimate the probability of the distribution of the maximum wind speed based on the error distribution between the above-mentioned point estimation of the maximum wind speed and the actual observed wind speed.
Step V. Calculate the upper limit of wind speed based on the probability distribution above (Table 7.5).

Table 7.5 Japanese rules for the operation management in crosswinds (wind speed refers to the instantaneous wind speed)

Wind speed /(m/s)	A certain interval	Windbreak wall interval with a certain standard
20 ≤ Wind speed < 25	The speed limit is below 160 km/h	No speed limit
25 ≤ Wind speed < 30	The speed limit of trains below 70 km/h is also subject to specific circumstances	The speed limit is below 160 km/h
30 ≤ Wind speed < 35	Stop running	The speed limit is below 70 km/h
Wind speed ≥ 35	Stop running	Stop running

7.4.3 Crosswind Operation Mechanism of China's HSR

China now makes use of information technology to execute online monitoring, monitor potential dangers and traffic safety during catastrophe events, and forecast or provide overrun warnings in accordance with the established threshold classification. China's crosswind warning signal regulations for the HSR safety operation are established in line with the General Technical Plan for Disaster Prevention and safety monitoring system of passenger dedicated lines (Interim). Except for typhoon-type crosswind, specific portions of the southeast coast, and crosswind sections, China's crosswind warning signals for the HSR safety operation are separated into four levels and expressed in blue, yellow, orange and red, as indicated in Table 7.6.

Based on Table 7.6, in the HSR safety operation, the crosswind warning signals for the HSR safety operation are:

① The blue warning signal is the first-level crosswind warning signal: When the instantaneous wind speed is 15.0 and 20.0 m/s, and the operating speed is controlled at 300 km/h.

② The yellow warning signal is the second-level crosswind warning signal: When the instantaneous wind speed of 20.0 and 25.0 m/s, the operating speed control is controlled at 200 km/h.

③ The orange warning signal is the third-level crosswind warning signal: when the instantaneous wind speed of 25.0 and 30.0 m/s, the operating speed control is controlled at 120 km/h.

Table 7.6 China high-speed railway operation limit standard of strong wind

Wind speed /(m/s)	Safety signal		Speed limit
15.0–20.0 m/s	Level 1	Blue warning signal	Control the speed at 300 km/h
20.0–25.0 m/s	Level 2	Yellow warning signal	Control the speed at 200 km/h
25.0–30.0 m/s	Level 3	Orange warning signal	Control the speed at 120 km/h
> 30.0 m/s	Level 4	Red warning signal	Stop running

④ The red warning signal is the fourth-level crosswind warning signal: m/s the EMU is taken out of service when the instantaneous wind speed is more than 30.0.

In the HSR safety operation, the red warning signal of special sections along China's southeast coastal and the national HSR crosswind sections (extra-large bridges, passes, canyons, wind inlets in mountain areas, gorge tube effect, etc.) is that when the instantaneous wind speed is greater than 25.0 m/s, the HSR will be shutdown. This is mainly due to the effect in wind speed and height of the HSR passing through southeast coastal special sections and the HSR crosswind sections. The influence of the topography increases the instantaneous wind speed by 1.23 to 1.7 times. If the included angle between the line and the strong wind direction is between 80° and 100°, the aerodynamic force of high-speed train will increase significantly under the influence of the instantaneous wind speed and crosswind resultant force under the strong wind weather conditions, and the possibility of train derailment and overturning will increase significantly. Therefore, a crosswind warning system must be established in order to guarantee the efficiency and safety of high-speed trains, as well as to reduce the time required for alarming and disarming.

7.5 Crosswind Warning System for the HSR Safety Operation

The crosswind warning system for the HSR safety operation is a subsystem of the disaster prevention and warning system, which consists of the monitoring units of the communication station, disaster prevention and safety monitoring center, transmission network equipment and other components and reserves the interfaces of other subsystems such as earthquake monitoring subsystem and icing monitoring subsystem. The system monitors the situation of wind direction, wind speed, rain and snow, foreign object intrusion, etc., at each monitoring point along the railway line of the passenger dedicated line, provides early warning information, discharges the early warning information, and issues traffic orders through the train control center, and issues rescue, maintenance, management and other orders through the disaster prevention center. The early warning process is shown in Fig. 7.9.

In the HSR safety operation, the process of the crosswind warning for the HSR safety operation is as follows.

Step 1: The crosswind parameter information monitoring module obtains relevant information about the environment where the acquisition device is located.
Step 2: The crosswind warning's parameter control module retrieves the most recent crosswind parameter data in accordance with the current time.
Step 3: The crosswind warning's information conversion control module will convert the obtained data information to AD.
Step 4: The data-aware control module of the crosswind warning calls the evaluation warning module to analyze the crosswind parameters after preprocessing. In

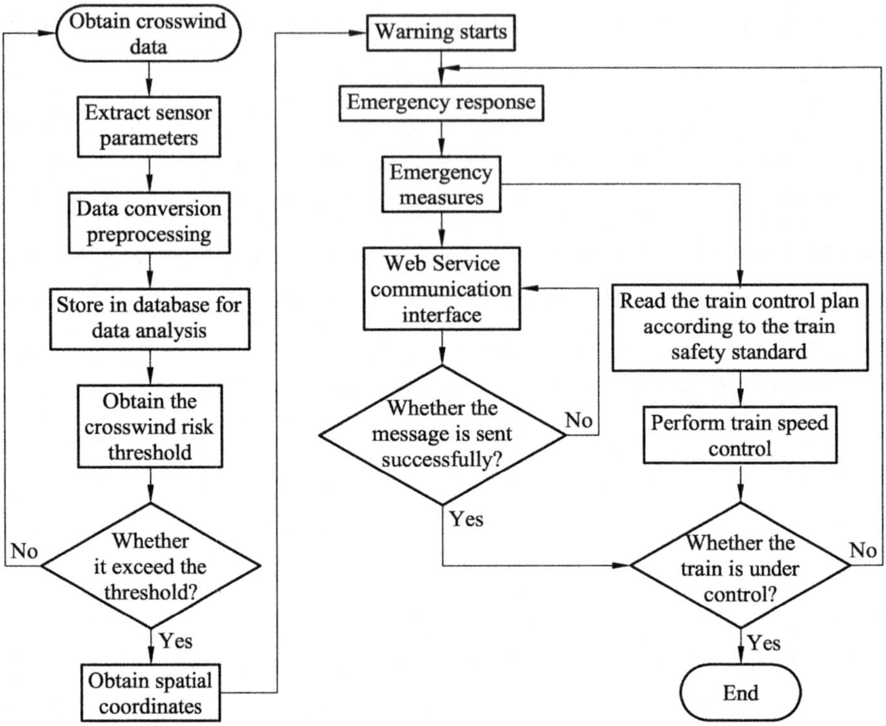

Fig. 7.9 Flowchart of the wind speed early warning system

order to determine the dynamic risk threshold, the crosswind parameter analysis calculates the average value of the current crosswind parameters as well as the fluctuation range of freshly obtained crosswind values.

Step 5: The threshold control module of the crosswind warning reads the risk threshold data of railway crosswind factors obtained after the analysis of the evaluation and warning module.

Step 6: The threshold interactive control module of the crosswind warning compares the safety risk threshold data of railway crosswind factors with the fluctuation interval value of crosswind parameters obtained in step 4 above.

Step 7: Return to step 2 and reread the crosswind parameter information if the line environment's crosswind information does not exceed the danger threshold. Step 8 should be taken if the line environment's safety information exceeds the danger level.

Step 8: Obtain the spatial coordinate information built into the control module to locate the affected lines.

Step 9: After positioning, the control module initiates the warning instruction and calls the warning process in the evaluation and warning module.

Step 10: Carry out emergency response according to the warning instruction.

Step 11: The control module puts into action the pre-established emergency plan, which primarily entails delivering the warning information and controlling train speed. The two programs are run concurrently, with steps 12 and 13 completing train control and steps 14 and 15 completing warning information transmission.
Step 12: Determine the relevant emergency filing measures for the railway line based on the safety level findings of the line crosswind factors obtained from the corresponding crosswind factor safety evaluation.
Step 13: Take train control emergency measures. Informing the train to slow down, halt, remove people to safe locations, etc., are the essential components.
Step 14: Send text-based emergency action information using a wireless transmission module or a mobile communication interface.
Step 15: Verify that the emergency measure's SMS message was properly sent: If yes, enter step 16; if not, (e.g., if the message is not sent successfully due to interference or other reasons) restart the communication module to send the text message of the emergency measure.
Step 16: Identify if the train control module has regulated the high-speed train's speed. If yes, enter step 17. If no (e.g., the train speed has not been controlled due to interference or other reasons), return to step 10 and repeat the warning measures.
Step 17: End the warning.

7.5.1 *Crosswind Warning System for the HSR Safety Operation*

There is a link between train speed and train load, which is one of the key influences of crosswind variables on the HSR in terms of safety operation. The use of light materials, which increases train speed, makes the problem of lateral strong wind more sensitive. On the other hand, the high-speed operation will generate buoyancy and pitching moment, and as the operating speed increases, the buoyancy and pitching moment increase as well, resulting in the train "floating." Trains running in this state are easily affected by crosswinds and cause traffic accidents. This chapter focuses on crosswind disaster prediction using disaster monitoring data, establishes a real-time monitoring network, offers scientific early warning data and achieves the goal of reducing disaster damage and ultimately ensuring driving safety.

(1) **Early warning steps for the crosswind disaster.** The HSR crosswind warning system plays a significant role in encouraging the HSR's growth in China. In the HSR safety operation, the early warning system can obtain observation data in real time and enhance the reliability, accuracy and real time of the observation data, so that the data can be transmitted to the monitoring center in a timely manner to improve the management ability of the dispatching department. The aerodynamic study of the crosswind and the wheel-rail dynamics analysis is necessary to ascertain the link between crosswind speed and safety limitations of the train speed.

Step 1. Determine the wind scenario during the operation of high-speed trains.
Step 2. Determine the aerodynamic performance through wind tunnel experiments or numerical simulations.
Step 3. Establish the multi-vehicle dynamics simulation model of the high-speed train.
Step 4. Calculate the wheel-rail dynamics parameters of the high-speed train and analyze the relationship between these parameters and the crosswind wind speed and vehicle speed.
Step 5. Determine the specification-required link between crosswind speed and vehicle speed.

(2) **Warning contents of the crosswind disaster**. Currently, an increasing number of wind speed and wind direction monitoring stations are being erected along the HSR, resulting in a vast amount of data information. To evaluate and handle these massive volumes of data in a timely, effective and efficient manner, an information system that can monitor and anticipate crosswinds is required, in order to assure the safety of high-speed trains running in the related environment. The two main data in crosswind information are the wind speed data and wind direction data. The wind is generated by the airflow caused by the difference in the atmospheric pressure. Both the wind speed and direction are periodically shifting. Therefore, volatility, intermittency and unpredictability are characteristics of the wind data.

① **Prediction of the HSR wind speed data**: The wind speed is the basis for the wind classification, and the unit is generally m/s or km/h. In general, the greater the wind speed, the higher the wind level, the more destructive the wind. The conversion of the wind speed and wind power is shown in Table 7.7.

② **Prediction of the HSR wind direction data**: The wind direction is another crucial piece of information about the crosswind when it comes to HSR safety operations. Due to the asymmetry of the flow field on the windward and leeward sides of high-speed trains, the vehicle will inevitably encounter the effects of

Table 7.7 Relationship between wind force and wind speed

Wind power	Wind speed /(m/s)	Wind power	Wind speed /(m/s)
0	0–0.2	9	20.8–24.4
1	0.3–1.5	10	24.5–28.4
2	1.6–3.3	11	28.5–32.6
3	3.4–5.4	12	32.7–36.9
4	5.5–7.9	13	37.0–41.4
5	8.0–10.7	14	41.5–46.1
6	10.8–13.8	15	46.2–50.9
7	13.9–17.1	16	51.0–56.0
8	17.2–20.7	17	56.1–61.2

transverse aerodynamic force and aerodynamic moment when the wind direction is at a specific deflection angle with the car body.

Crosswind warning is a long-term process in the HSR safety operation, and there are many wind measuring stations along the HSR, so the accumulated monitoring data are huge. As a result, a huge quantity of data must be analyzed and processed in order to integrate the crosswind distribution characteristics with the parameter information along the HSR line and identify the critical places for implementing crosswind speed prediction in the crosswind monitoring and warning system.

7.5.2 Data-Aware Model for the Crosswind Warning

Data processing in the HSR safety operation primarily uses monitoring data to anticipate wind speed in the upcoming forecast cycle. To fulfill the system's needs for crosswind data prediction, particularly the high precision prediction requirement, the Kalman filter theory is used to evaluate and forecast the HSR crosswind data. The research's optimal prediction is applied to the crosswind monitoring and warning system, resulting in a system with crosswind prediction functions such as real-time modeling, real-time prediction and output, and model tracking and correction.

(1) **Working principle of the Kalman filter**. Kalman filter is an advanced control method and a prediction method based on linear regression analysis. The Kalman filter algorithm is a finite calculation method for estimating the system state with the minimum mean square error, i.e., by feeding the forecast error of the previous moment into the original forecast equation; the coefficients of the forecast equation are corrected in time to improve the forecast accuracy of the next moment. The working process of the Kalman filter is shown in Fig. 7.10.

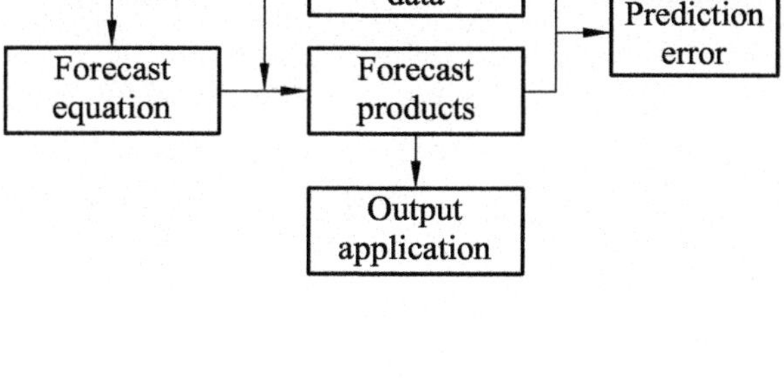

Fig. 7.10 Working process of the Kalman filter

(2) **Kalman filter estimation model for the crosswind warning**. In the HSR safety operation, the Kalman filter estimation model for the crosswind warning mainly consists of two parts: Kalman filter estimation (divided into prediction and filtering) and the calculation of Kalman (gain and estimation error) variance matrix. The process is as follows.

Step 1. In the HSR safety operation, the predicted value of the crosswind warning is calculated based on the previous filtered value.

$$\overline{X}(k|k-1) = \Phi(k,k-1)\overline{X}(k-1|k-1) \tag{7.6}$$

Step 2. In the HSR safety operation, the filter estimation is calculated from the new observations of the crosswind warning.

$$\overline{X}(k|k) = \overline{X}(k|k-1) + K(k)\big[z(k) - H(x)\overline{X}(k|k-1)\big] \tag{7.7}$$

Step 3. Store the filtering estimation into the computer and repeat the above calculation process when more crosswind warning observations become available.

Step 4. Define the Kalman gain function and the filter error variance matrix for the crosswind warning.

In the HSR safety operation, the Kalman gain function of the crosswind warning is:

$$K(k) = P(k|k-1)H^T(k)\big[H(k)P(k|k-1)H^T(k) + R(k)\big]. \tag{7.8}$$

In the HSR safety operation, the filter error variance matrix for the crosswind warning is:

$$P(k|k) = [1 - K(k)H(k)]P(k|k-1). \tag{7.9}$$

(3) **State-space model of the crosswind warning.** In the HSR safety operation, the state-space model for the crosswind warning is solved as follows:

Step 1. Determine the coefficient matrix, the initial state value and the variance matrix of the initialized system of the crosswind warning.

Step 2. State recursive calculation of the crosswind warning system state is as follows:

$$x_{n|n-1} = F_n x_{n-1|n-1} \tag{7.10}$$

$$V_{n|n-1} = F_n V_{n-1|n-1} F_n^T + G_n Q_n G_n^T, \tag{7.11}$$

where

$x_{n-1|n-1}$ is the filtered value of the crosswind warning system state during the period $n-1$;

$x_{n|n-1}$ is the state vector of the crosswind early warning system during the predicted period n according to the filter value of the crosswind early warning during the period $n-1$;

$V_{n-1|n-1}$ is the state variance matrix of the crosswind warning during the period $n-1$;

$V_{n|n-1}$ is the state variance matrix of the crosswind warning system during the predicted period n based on the state variance matrix of the crosswind warning during the period $n-1$.

Step 3. Modify the state equation of the crosswind warning system according to the observation vector filtering:

$$K_n = V_{n|n-1} H_n^T \left(H_n V_{n|n-1} H_n^T + R_n\right)^{-1} \tag{7.12}$$

$$x_{n|n} = x_{n|n-1} + K_n\left(y_n - H_n x_{n|n-1}\right) \tag{7.13}$$

$$V_{n|n} = (I - K_n H_n) V_{n|n-1}, \tag{7.14}$$

where K_n is the Kalman gain matrix for the crosswind warning;

$x_{n|n}$ is the filtered value of the system status of the crosswind warning during the period n;

$V_{n|n}$ is the modified value of the state variance matrix of the crosswind warning system during the period n.

In the HSR safety operation, the initial value of the system state x_0 of the crosswind warning and the variance matrix of the crosswind warning V_0, $V_0 = D(x_0)$ are known. For $n = 1$, if x_0 is substituted into Eq. (7.10), $x_{1|0}$ can be obtained; if V_0 is substituted into Eq. (7.11), $V_{1|0}$ can be obtained; if K_1 is substituted into Eq. (7.12), K_1 can be obtained; if K_1 is substituted into Eq. (7.13), $x_{1|1}$ can be obtained. In this way, from x_0 and the crosswind warning observation value y_1 at the time of $n = 1$, the optimal estimation value $x_{1|1}$ of the crosswind early warning system at the time of $n = 1$ can be obtained by using Eq. (7.14).

In order to perform the following recursion, K_1 needs to be substituted into Eq. (7.14) to get $V_{1|1}$, so as to get $x_{1|1}$ and $V_{1|1}$. For $n = 2$, continue to calculate according to the above steps to obtain $x_{2|2}$ and $V_{2|2}$. By analogy, we can obtain the optimal estimation and variance matrix of the crosswind early warning when $n = 3, 4, 5$.

7.5.3 Early Warning System for Crosswind Disaster Monitoring

The present railway crosswind monitoring strategy is to place wind speed and direction sensors and acquisition units along the HSR line and gather wind speed and direction data in real time. An alarm will be generated if the data exceed the alarm value. After the user confirms the alarm information and site conditions, he/she shall take timely response measures such as slow down parking or avoidance. The users of the disaster prevention system are generally railway dispatchers. If severe winds, particularly significant crosswinds, are notified in advance before they disrupt regular line operation, dispatchers will have more time to devise acceptable and adequate remedies. The HSR crosswind monitoring and prediction system suited for China's national circumstances is designed based on a comparison of domestic and international railway crosswind monitoring and early warning systems, and it includes:

① The design of the information gathering interface for wind speed and wind direction is achieved using the wind speed and wind direction sensors.
② The wireless transmitter module for 802.11-WIFI enables the real-time transmission of sensor data.
③ The higher computer's data fusion processing center is created using the Visual Studio platform.
④ On the basis of traditional wind speed early warning and discrimination, considering the influence of wind direction on train operation, an assessment algorithm for the HSR safety operation based on wind direction and wind speed is designed.
⑤ The information pushing function for the HSR safety operation assessment is accomplished via the mobile communication base station.

The four distinct components of the design process for the wind speed detection-based HSR warning system are threshold discrimination, warning algorithm, system structure and integration test.

7.5.3.1 Early Warning Threshold Definition

The limit wind speed of overturning, which is often assessed by the limit wind speed of overturning, is affected by wind direction, wind speed, vehicle weight, form, track subgrade construction and high-speed train operating speed (the minimum natural wind speed to ensure the HSR safety operation when other factors remain unchanged). After the train operating chart is established, the speed, subgrade structure and other parameters have been fixed, the limit wind speed of overturning can also be calculated through simulation. According to the current national established standards, the train operates normally when the ambient wind speed surrounding the railway line is larger than 15 m/s; however, when the crosswind speed along the

Table 7.8 Discriminative threshold of wind speed along railway

Ambient wind speed along the railway (m/s)	Speed limit for train operation
[0, 15]	Normal operation
[15, 20]	The speed limit of 300 km/h
[20, 25]	The speed limit of 200 km/h
[25, 30]	The speed limit of 300 km/h
> 30	No train is allowed to enter the wind zone

railway line is greater than 30 m/s, the train comes to a complete stop. The specific discriminatory thresholds are shown in Table 7.8.

7.5.3.2 Dual Factor Algorithm of the Wind Speed and Wind Direction

High-speed trains are not affected by wind direction, as the standard crosswind warning method for the HSR is based on wind speed. To address this problem, a two-factor discrimination algorithm based on wind speed and wind direction is proposed, assuming that the direction of the train travel is N, the speed of the crosswind is v_0(m/s), and the wind direction is N + 90° + a, as shown in Fig. 7.11.

In the HSR safety operation, the projection vector of the crosswind in the vertical direction of train operation is

$$V_1 = V_0 \times \cos a. \tag{7.15}$$

In the HSR safety operation, the projection vector of the crosswind in the parallel direction of train operation is

$$V_2 = V_0 \times \sin a. \tag{7.16}$$

If the parallel direction of the crosswind is the same as the forward direction of the train, this portion of the crosswind vector will not interfere with the HSR operation; if the parallel direction of the crosswind vector is opposite to the forward

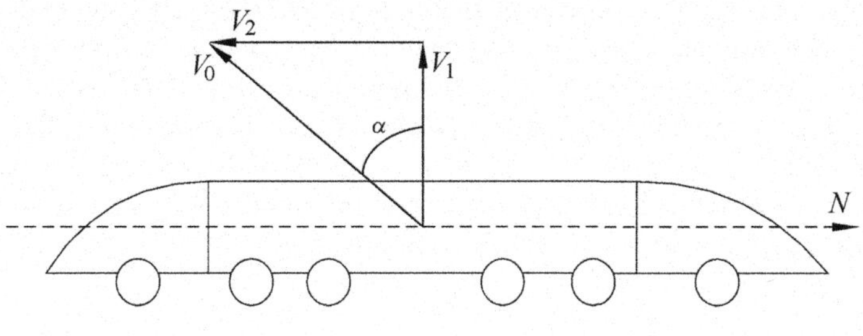

Fig. 7.11 Schematic diagram of crosswind interference in high-speed railway vehicles

Table 7.9 Hardware alarm mode of the wind speed along the HSR

Wind speed level	Alarm mode	Intelligent operation mode	Operating mode
Level 1	The buzzer does not alarm	Send text message	Order the trains to run normally
Level 2	Buzzer alarm (intermittent)	Send text message	Order the trains to run normally
Level 3	Buzzer alarm (rapid)	Send text message	Order trains to slow down

direction of the train, it will produce large resistance and centrifugal force to the HSR operation. Therefore, the early warning discrimination algorithm for the HSR based on the wind direction and wind speed is: Early warning discrimination solely relies on the projection of the vertical direction of crosswind if the parallel direction of crosswind and the forward direction of the train are the same; assuming that the parallel direction of crosswind is opposite to the forward direction of the train, the early warning judgment is made according to the sum of the projection vector in the vertical direction of crosswind and the projection vector in the parallel direction.

$$V = \begin{cases} v_1, & 0^\circ < a < 180^\circ \\ v_1 + v_2, & 180^\circ < a < 360^\circ \end{cases}, \tag{7.17}$$

where v is the final threshold discriminating wind speed, m/s.

In the specific design, the following methods are used for hardware alarm; see Table 7.9.

7.5.3.3 Early Warning Architecture of the Crosswind Monitoring System

An upper computer, a lower computer, and a terminal comprise the HSR warning system based on wind speed discrimination. Among them, the upper computer includes wind speed and wind direction sensor which collect the wind speed and wind direction, and the central processor which controls the work of the wind speed and wind direction sensor. The HSR warning system's top computer serves as the simulation platform and is used to read and dynamically process data input from the central processor through the wireless module in real time, such as including data decoding, warning and discrimination. The terminal is the final information display object of the early warning system, and it may convey the railway early warning information to the mobile phone interface via information push for passengers, drivers or railway operation management departments (Fig. 7.12).

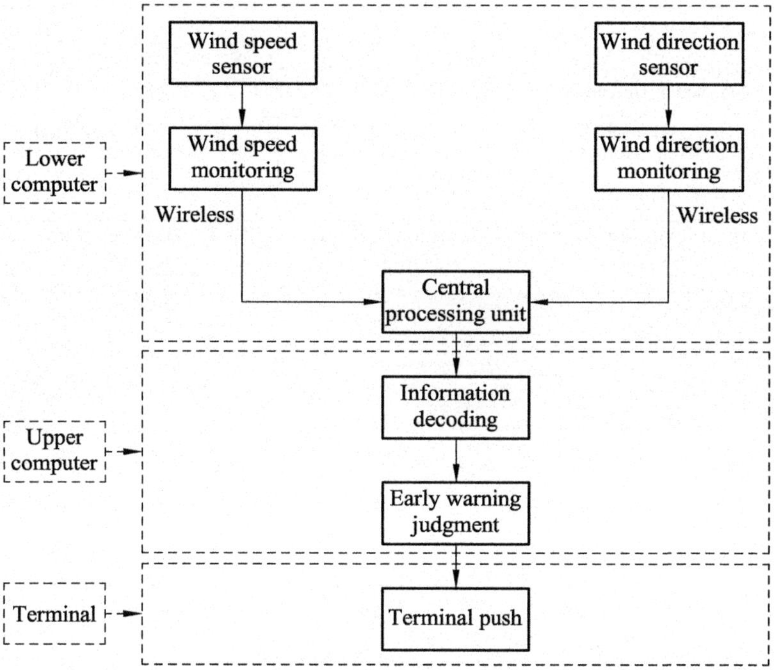

Fig. 7.12 Framework of the HSR early warning system based on the wind detection

7.5.3.4 Test Analysis of the Crosswind Monitoring System

The HSR warning system based on the wind speed detection was tested using a simulated crosswind to ensure the stability and viability of the system. The test method is to first load a small wind speed at a certain angle to the wind speed sensor, and then change the wind speed loading angle and increase the wind speed on this basis to test the changes of the wind speed information, wind direction information and warning information collected by the whole system in the process of wind speed change. Due to resource limits, it was sometimes difficult to simulate wind speeds beyond 20 m/s during the real test. Therefore, the early warning level is divided into three levels on the basis of the national standard wind speed early warning interval in order to highlight the sensitivity and practicability of the system. The final test results are shown in Fig. 7.13.

The test results in Fig. 7.13 show that the crosswind early warning system can quickly and accurately collect the current wind speed and direction information, make early warning judgments on the safety of the HSR operation based on the current information and push the information to the mobile terminal.

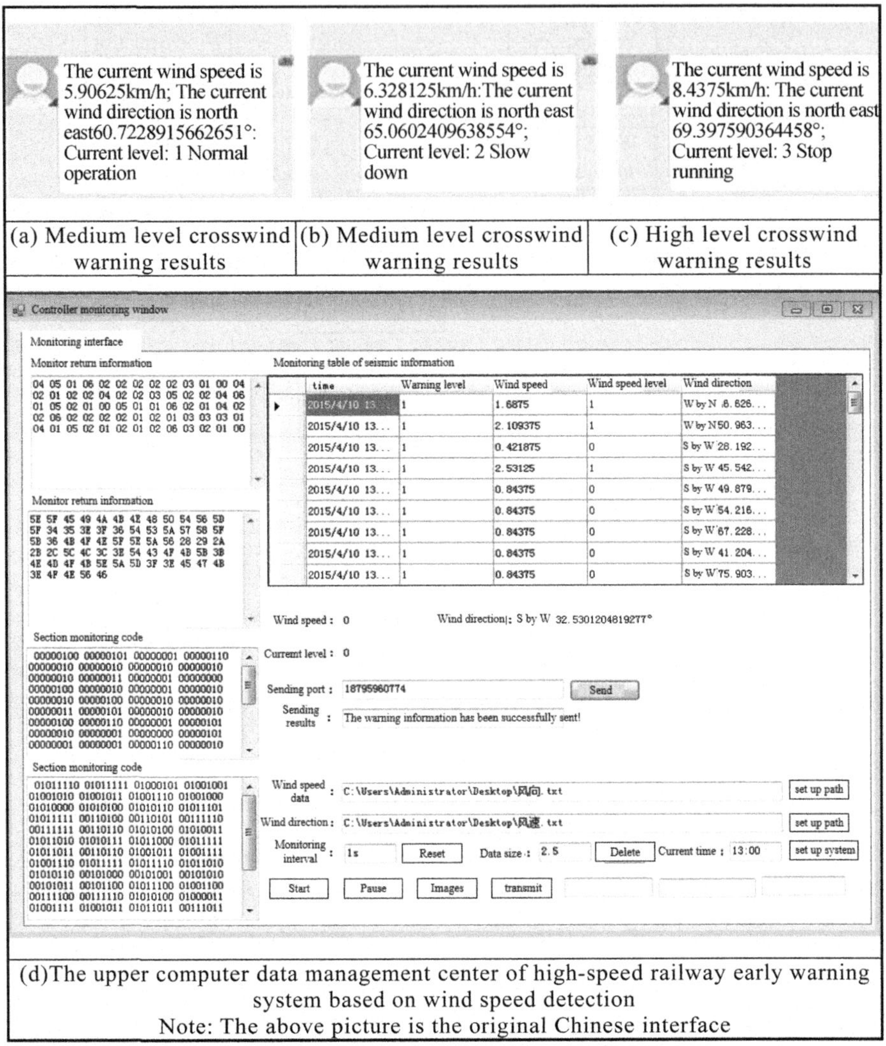

Fig. 7.13 Test results of high-speed railway early warning system based on wind speed detection

7.6 Crosswind Warning System for the HSR Safety Operation

This chapter builds a crosswind warning system for the HSR safety operation in order to assure HSR safety operation, particularly in crosswind circumstances.

7.6.1 *Architecture of the Crosswind Warning System for the HSR Safety Operation*

Through the comparative analysis of the HSR crosswind monitoring and early warning system at home and abroad, the HSR crosswind monitoring and early warning system can be designed as a four-layer structure, namely, the field layer, remote terminal layer, central processing layer and application layer. The crosswind warning system for the HSR safety operation is built in four tiers.

7.6.1.1 Field Layer of the Crosswind Warning System

The sensor layer of the crosswind warning system is the field layer in the HSR safety operation, which includes the installation of the aerovane and supporting transmission equipment. The field layer of the crosswind warning system, which is the foundation of the entire system, may immediately acquire wind speed and wind direction information along the track. Three issues need to be considered in the design of the site layer:

(1) **Selection of the aerovane for the crosswind warning system**. The aerovane mainly includes the three-cup aerovane and propeller aerovane, acoustic aerovane and thermal field aerovane. Among them, the three-cup and propeller aerovanes, while inexpensive, have substantial maintenance costs and are unsuitable for high dependability requirements of the HSR. The properties of ultrasonic and thermal field aerovanes include high precision, high reliability and high-cost performance, which can meet the requirements of the HSR. The ultrasonic aerovane needs to be heated when used in cold environments, otherwise, the probe icing or frosting will affect the measurement results.

In the HSR safety operation, due to the complex meteorological conditions along the HSR, the thermal field aerovane is typically employed.

(2) **Distribution of the aerovane for the crosswind warning system.** There are two main kinds of distribution methods: the key location method and the covering method. Among them, the first method is the key location method, which involves using the wind speed and direction data from meteorological stations along the railway for many years, combined with the field topographic survey and geographical parameters along the railway, to perform the time distance conversion of the maximum wind speed, establish the probability model of the maximum wind speed, obtain the maximum wind speed distribution curve of the whole line and finally determine the layout plausibility to meet the needs of the project planning, design, construction and safe operation for passenger dedicated lines. The second method is the covering method: in which the air flow model of the section is finished using computational fluid dynamics (CFD) technology in accordance with the topographical characteristics, surface conditions, embankment and viaduct settings of the supplied line. Then, based on

the model, curve and inclination data of the line itself, and aerodynamic data of running vehicles, the conversion coefficient between the average wind speed and the most unfavorable wind direction at each 50-m interval along the line and the measured data at the wind measuring point is calculated to form the relationship curve between the dangerous wind speed and wind direction in this section, that is, the critical curve.

In the HSR safety operation, both methods can achieve the goal of deployment, of which the covering method is more operational since there is no need to worry about the omission of deployment points.

(3) **Installation of the aerovane for the crosswind warning system**. There are two ways to install the aerovane. One of them is to mount a wind speed and direction indication on a catenary pillar next to the line. Another is to mount an aerovane on the GSM-R tower away from the line. For the crosswind warning and monitoring system, these two installation options have their own advantages and disadvantages. The advantage of mounting the aerovane on the GSM-R tower is that it is close to the base station and convenient to lay the transmission cable, but the downside is that the position with the tower is not always the site where the wind speed and direction meter should be installed. The aerovane can better fit the requirements for installation in the crosswind early warning and monitoring system if it is positioned on the catenary pillar. However, the majority of HSR base stations are located beneath the bridge, and the transmission cable laying requires up-and-down bridge protection. To ensure the accuracy and reliability of the data monitored by the aerovane, the aerovane is uniformly set on the catenary mast.

According to the study the optimum wind speed and direction data for HSR safety operations come from the aerovane positioned on the catenary pole at a height of 4 m above the rail surface.

7.6.1.2 Remote Terminal Layer for the Crosswind Warning Systems

The remote terminal layer for the HSR safety operation is primarily in charge of monitoring the operational conditions of various elements of the system, which is composed of different monitoring units. The monitoring unit equipment of the crosswind warning system is installed in the GSM-R base station near the on-site detection equipment and placed in parallel with other communication cabinets. The monitoring unit for the HSR safety operation's primary responsibilities includes gathering real-time data from the on-site monitoring equipment (typically via RS-232 or RS-485 connection between the remote terminal layer and the on-site layer, which is primarily determined by the aerovane used), processing and temporarily storing of the data by the embedded acquisition software, and then uploading to the central processing layer via the transmission network. It also monitors and manages the status information of the on-site monitoring equipment, conducts self-inspection, enables fault alarm, fault diagnosis and fault location, uploads fault records and other information and accepts centralized monitoring and management of the central processing layer. The design of this layer includes two issues.

(1) **Data transmission between the remote terminal layer and the central processing layer of the crosswind warning system**: wired or wireless mode. The first is the wired mode. For the crosswind monitoring and early warning system that has been built at present, the unique channel is utilized between the remote terminal layer and the central processing layer, which ensures the real time and efficiency of data, but the construction cost is high, and the line is not conducive to maintenance. The second is wireless mode, which is based on wireless network design. At present, there are three main ways to rent wireless public networks for data transmission: ① SMS-based data transmission; ② GPRS-based data transmission; ③ CDMA-based data transmission. For industrial data, which requires transmission speed and reliable transmission, CDMA is generally used for transmission. GPRS transmission can be taken into consideration if the quality and speed of transmission are impacted by inadequate CDMA network coverage circumstances; the comparatively straightforward GSM/SMS implementation approach can be taken into consideration if the transmission is utilized for civilian purposes if the speed and quality requirements are not great. However, if a GSM-R network is built along the railway line, the GPRS packet-switched data transmission mode is introduced accordingly.

In the HSR safety operation, considering the requirements of data transmission speed, transmission reliability, complexity of technology implementation, and network coverage, GSM-R/GPRS is the optimal for transmitting data.

(2) **Practicality and reliability of the monitoring unit of the crosswind warning system**: Long-term engineering practice proves that the reliability of an unattended device must include the following four aspects: (i) quality control of the device itself; (ii) quality control of the third-party peripherals used by the device; (iii) quality control of the communication link; and (iv) remote monitoring and manual intervention of the device status.

The first three components of the HSR safety operation demand that the system be capable of automatic detection and recovery, while the last component ensures that the system will recover even under the most challenging circumstances. For the crosswind warning system monitoring unit to comply with system dependability criteria, it must fulfill the aforementioned four requirements.

7.6.1.3 Central Processing Layer of the Crosswind Warning System

The central processing layer of the HSR safety operation is in charge of receiving various information transmitted from each monitoring and processing layer within the jurisdiction in real time, storing, analyzing and processing data in real time and providing disaster prevention alarms and warnings of corresponding levels in accordance with the content of the information. It also gives information on speed limitations, stoppages and other plans in accordance with train operation control standards and uploads the alarm information to the users of the application layer. This layer mainly consists of database server and application server. The database server is a hot standby mode. It is outfitted with corresponding monitoring data tables to keep track of different system data, such as wind direction and speed information, alarm information, equipment fault information and other historical records for query. The core issue of the design of this layer is to use the actual measured wind speed and wind direction data for early warning. The research on the crosswind early warning for the railway is still in its infancy. On the basis of several existing early warning methods, the model, algorithm, theoretical basis and practical application of early warning must be empirically researched and validated.

The present monitoring of crosswind in the HSR safety operation is to set up monitoring stations along the HSR, install wind speed and wind direction sensors and collecting units, and collect wind speed and wind direction data on a timely basis. An alert will sound if the data reach the alarm value; after verifying the warning information and the situation on the scene, he/she shall take timely countermeasures, such as slowing down, stopping or avoiding. The users of the disaster prevention system are generally railway dispatchers. If strong winds, especially strong crosswinds, are warned in advance before crosswinds affect the normal operation, dispatchers will have more time to devise acceptable and adequate countermeasures, which will play an important role in averting disasters and guaranteeing traffic safety and efficiency. In the HSR safety operation, the early warning of strong wind can be divided into three methods: time point early warning, spatial point early warning and alarm value warning.

The first method: time point warning method. The time point warning refers to that all monitoring points of the whole line have a separate predicted value at some point in the future. For example, the JR-EAST's crosswind warning system.

The second method: spatial point warning method. The spatial point warning is aimed at the whole line; that is, the system receives data from the meteorological department, including wind force and wind direction changes, etc. If a crosswind situation exists, the system determines whether the crosswind has an effect on the railway line and notifies the user of the computed findings, for example, the Japan's WINDAS system.

The third method: alarm value warning method. The alarm value warning is a comprehensive early warning for a single monitoring point and the whole line, combined with meteorological, train, line and other conditions. In the design of the crosswind warning, the user's needs and railway management regulations are taken into consideration. Once a single monitoring point hits the alert value, alarming is not permitted. After the user confirms, the alert will not be repeated or the number of alarms won't be decreased during the length of the heavy wind. Early warning and alarm also follow the same alarm principle. Such as Qinghai–Tibet Railway crosswind monitoring system.

7.6.1.4 Application Layer of the Crosswind Warning System

In the HSR safety operation, the application layer of the crosswind warning system offers data display, printing and other functions to help users make informed decisions about disaster prevention. It also offers interfaces to the train control system so that trains traveling through windy regions can be controlled by timely speed limits or stoppages.

7.6.2 Design of the Remote Terminal Monitoring Unit for the Crosswind Warning System

In the HSR safety operation, the main task of the monitoring unit for the HSR safety operation is to collect, store and transmit wind direction and wind speed data, and its structure design is shown in Fig. 7.14.

(1) **Signal conditioning for the crosswind warning system**: In the HSR safety operation, the operation amplifier and filter circuit are added at the front end of the system to improve the received signal's accuracy and make it simpler to calculate and process because the signals provided by the wind direction and speed sensor are typically weak or contain high-frequency noise.
(2) **Acquisition and storage of the crosswind warning system**: Since a monitoring unit needs to acquire multiple signals, and the cost of AD chips for the multiple acquisition is usually high, we can consider using analog switch chips and the single-acquisition AD chips to reduce the implementation cost. On the other hand, because the data sampling rate is generally high and the data transfer rate between the distant terminal layer and the central processing layer is restricted,

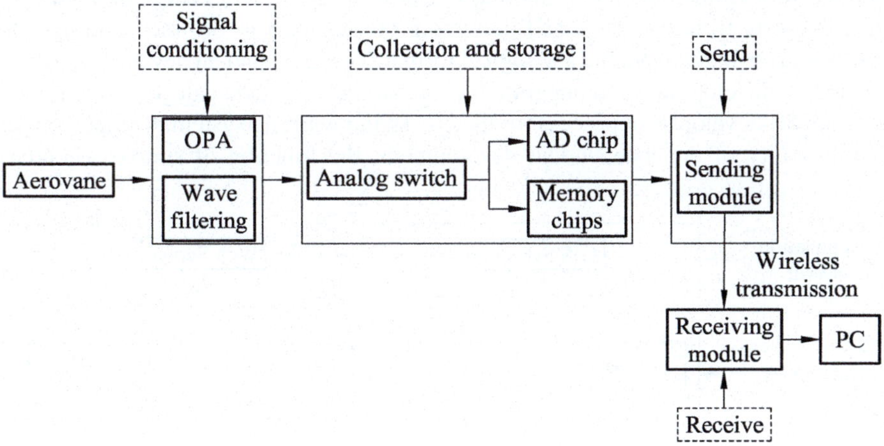

Fig. 7.14 Structure design drawing of the monitoring unit

the collected data should be stored in the storage chip of the monitoring unit in the first place.

(3) **Information sending of the crosswind warning system**: In the HSR safety operation, it is feasible to reply to data requests from the central processing layer to the remote terminals as well as transfer the gathered data at regular intervals to the server. Due to the development of AD chips, many AD chips have built-in programming processors, such as THS1206, AD7725 and other chips. Therefore, the program processing of data can only rely on the AD chip itself, without the need to introduce additional single chips, due to the design expense and complexity of implementation.

7.6.3 *Software Design of the Crosswind Warning System*

Due to the long distance and large span of the railway lines, each station terminal or relevant department needs to know the local and along-line crosswind information in real time, which necessitates the system software must be convenient and easy to deploy and maintain in bulk, and the browser/server (B/S) structure is undoubtedly the best choice. This is due to the fact that the design and maintenance of the B/S structure software are centered on the server side, and each client may access the system data via a browser to grasp the real-time wind information along the railway line. The system software must be constructed using a geographic information system (GIS) since the monitoring data for the system is typically tightly tied to geographic coordinates and terrain topography. In the HSR safety operation, the software architecture of the crosswind warning system is shown in Fig. 7.15.

As can be seen in Fig. 7.15, the geographic information server and the web server form the core of the crosswind warning system software. The GIS server of the crosswind warning system manages the system data such as wind speed and direction and the geographic data including maps. The web server is in charge of answering data requests from browsers and delivering the answers to browsers through the Internet in an understandable and logical format. The independent development, pure secondary development and integrated secondary development categories are typical for the GIS development process for the crosswind warning system. Compared with the difficulty of the independent development, the function of simple secondary

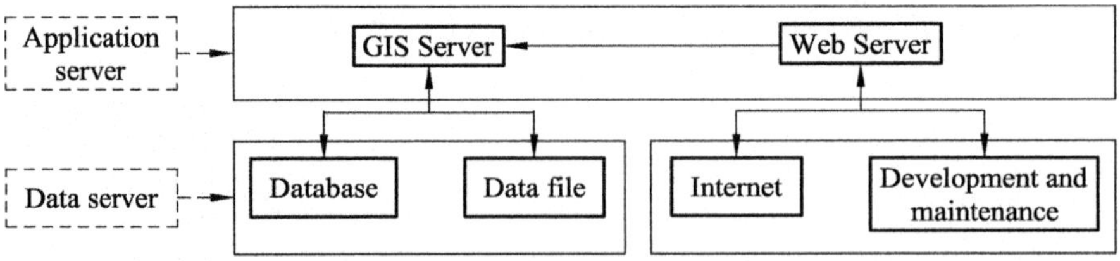

Fig. 7.15 System software architecture diagram

development is relatively simple. The professional GIS tools software will be used in the integrated secondary development to realize the fundamental GIS functions, and general software development tools, particularly visual development tools, will be used as the development platform to execute the integrated development of the two. This approach combines the benefits of the preceding two development methods by integrating sophisticated GIS functionalities as well as independent research and development for particular applications.

7.7 Summary of This Chapter

This chapter conducts a systematic study on the HSR safety operation under the crosswind disasters. Based on the analysis of domestic and foreign HSR crosswind monitoring systems, this chapter provides an early warning system for the HSR safety operation during crosswind catastrophes and uses remote monitoring units and geographic information systems as its basic components.

The crosswind early warning system uses continuous wind speed measurement data to predict the maximum possible wind speed (the maximum predicted wind speed) when trains travel over a certain segment in a crosswind weather. If either the maximum predicted wind speed or the actual measured value (i.e., the actual wind speed) exceeds the set value of the train control, a speed restriction or suspension of operation is implemented in a timely manner. However, only when both the maximum predicted wind speed and the actual wind speed are less than the control set value, can the control be released. This chapter develops a remote monitoring system for the HSR crosswind monitoring and early warning based on the GIS and sensor technologies after studying domestic and international railway crosswind safety monitoring systems.

Bibliography

1. Rungskunroch L, Panrawee AJ (2021) Bench marking on railway safety performance using Bayesian inference, decision tree and petri-net techniques based on long-term accidental data sets. Reliab Eng Syst Saf 213:107684
2. Shengen T, Xuebing L, Jiye Z, Weihua Z (2008) The Flow field struetuer and the aerodynamic performanee of high speed trains running on Embankment under cross wind. Rolling Stock 46(8):4–8
3. Mengge Y, Jiye Z, Weihua Z (2011) Wind-induced security of high-speed trains on the ground. J Southwest Jiaotong Univ 46(6):989–995
4. Pagliara F, Mauriello F (2020) Modelling the impact of high speed rail on tourists with geographically weighted poisson regression. Transp Res Part A Policy Pract 132(2):780–790
5. Hu Q, Fang X, Bian L (2021) Natural disaster warning system for safe operation ofa high-speed railway. Transp Saf Environ 3(4):1–12
6. Ren Zunsong X, Yugong WL, Yingzheng Q (2006) Study on the running safety of high-speed trains under strong cross winds. J China Railway Soc 28(6):46–50

7. Xuebing L, Zheng Y, Jiye Z, Weihua Z (2009) Aerodynamics properties of high-speed train in strong wind. J Traffic Transp Eng 9(2):66–73
8. Mi D (2018) The potential impact of high-speed rail on the economic geography of China. Transp Res Part A Policy Pract 113:279–290
9. Wang J, Rakha HA (2018) Longitudinal train dynamics model for a rail transit simulation system. Transp Res Part C: Emerg Technol 86(1):111–123

Chapter 8
An Integrated Natural Disaster Warning System for the HSR Safety Operation

For the HSR system, the safe operation is the top priority. To avoid potential natural catastrophes from disrupting the HSR's routine operation, the HSR should construct an integrated monitoring system for natural disaster warning. The system is utilized to thoroughly track the effects of natural catastrophes (such as wind, rain, lightning, temperature, debris flow, earthquake and other natural disasters) on the HSR safety operation. Based on an analysis of the impact of natural disasters on HSR safety operations, this chapter builds a comprehensive early warning system for HSR safety operations under natural disasters, which contains the early warning systems for the wind, rain, lightning, temperature, geology and earthquake, as shown in Fig. 8.1.

According to Fig. 8.1, in the HSR safety operation, the HSR integrated natural disaster warning and monitoring system can be generally divided into four levels.

Level 1: the user layer of the HSR integrated monitoring system. Dispatchers utilize the user layer of the HSR, and the equipment for this layer is housed at the HSR monitoring center. On the one hand, the monitoring center of the HSR serves the high-speed lines within the jurisdiction of the whole monitoring center; on the other hand, it provides the alarm information.

Level 2: the regional processing layer of the HSR integrated monitoring system. The HSR's regional processing layer is primarily in charge of receiving various information transmitted in real time from each monitoring and processing layer under its purview, storing, processing, analyzing, displaying and printing data and providing the appropriate level of monitoring and warning in accordance with the information content. In addition, according to the control rules of the high-speed train operation, it delivers the plan information such as speed limit and stoppage and uploads the alarm information to the user layer of the HSR.

Level 3: the monitoring and processing layer of the HSR integrated monitoring system. The HSR monitoring and processing layer collects real-time data from on-site monitoring equipment, processes and stores it for a brief period of time, and then uploads it to the HSR regional processing layer through the network. It can also

Q. Hu, *Natural Disaster Warning System for High-Speed Railway Safety Operation*, Advances in High-speed Rail Technology, https://doi.org/10.1007/978-981-99-7115-2_8

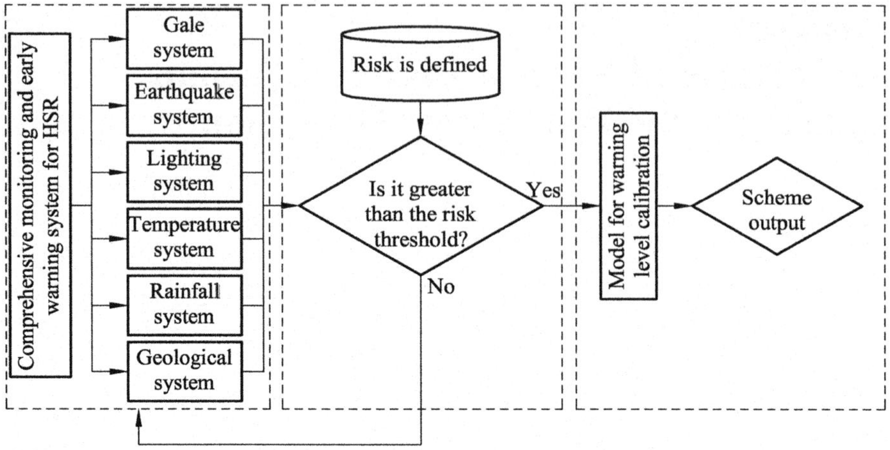

Fig. 8.1 An integrated monitoring system for the early warning of natural disasters

monitor and manage the status information of the on-site monitoring equipment to realize fault alarm and fault diagnosis.

Level 4: the on-site information acquisition layer of the HSR integrated monitoring system. The on-site information acquisition layer of the HSR is directly oriented to the trackside facilities. The HSR trackside facilities are not only scattered but also varied in substance, to enable the HSR's safety operation in the event of a natural calamity.

8.1 Integrated Natural Disaster Monitoring System for the HSR Safety Operation

China's high-speed railways have high speed (350 km/h) and high train density (the minimum train tracking interval can reach 3 min). Natural catastrophes continue to pose a serious danger to China's HSR safety operation despite the country's strong requirements for HSR development. According to the previous chapters, crosswind, temperature, precipitation, earthquake, lightning and geology are the primary natural calamities that endanger the safety of high-speed trains. Therefore, a comprehensive monitoring system for the HSR safety operation during natural disasters is built based on the building of the HSR monitoring subsystems for various natural disasters.

(1) **HSR crosswind warning system**. In the HSR safety operation, when the high-speed train is running on the HSR line, the dynamic parameters of the high-speed train, including the derailment coefficient, load reduction rate, overturning coefficient and wheel-rail lateral force, increase significantly under the action of the crosswind, thus reducing the safety and reliability of high-speed train

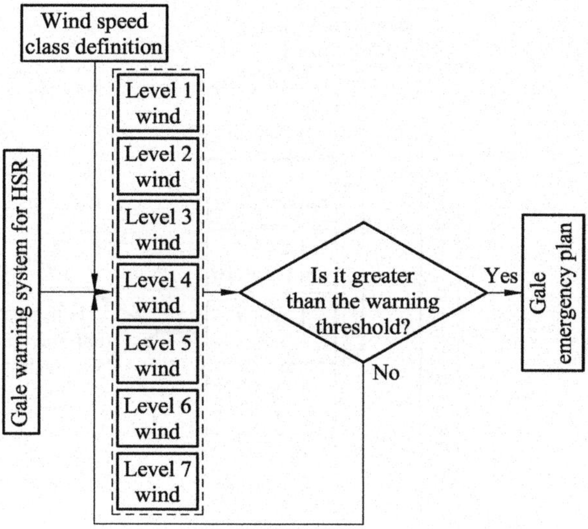

Fig. 8.2 HSR crosswind warning system

operation. Among them, the derailment coefficient and load reduction rate of high-speed trains running on curves are dramatically increased by the effect of the lateral wind on the inner side of curved tracks, which is a relatively risky working condition in the high-speed train operation. This chapter constructs the HSR crosswind warning system based on research findings both at home and abroad, as shown in Fig. 8.2.

(2) **HSR temperature warning system.** In the HSR safety operation, high temperature or low temperature will make the rail expansion or contraction. A rail temperature monitoring system is necessary for high-speed rail in sections where the ballast track's curve radius is less than 6000 m, at the ends of exceptionally large continuous bridges with a large temperature span, and in sections with numerous bridges, according to the relevant research findings of China's Ministry of Transport. This chapter constructs the HSR temperature warning system, as shown in Fig. 8.3.

Snowstorms in the northeast of China have a large impact on the HSR operation. Continuous snowfall and severe cold weather may cause excessive load on the contact network and power supply lines, preventing the switch from being converted. Therefore, the effective monitoring must be carried out. This chapter constructs a HSR snow disaster warning system based on domestic and foreign research results, as shown in Fig. 8.4.

(3) **HSR rainstorm warning system**. Continuous heavy rain or rainstorms might increase the risk of debris flows or subgrade collapses during HSR safety operations. It is typically unnecessary to take the passive monitoring alarm into consideration for ballastless tracks and high-speed trains with a high proportion of bridges and tunnels. According to the research results at home and abroad, the HSR rainstorm warning system is constructed, as shown in Fig. 8.5.

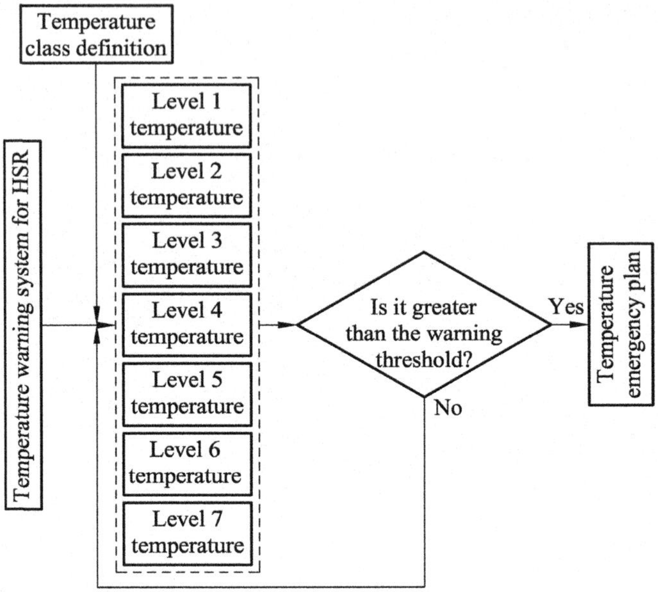

Fig. 8.3 HSR temperature warning system

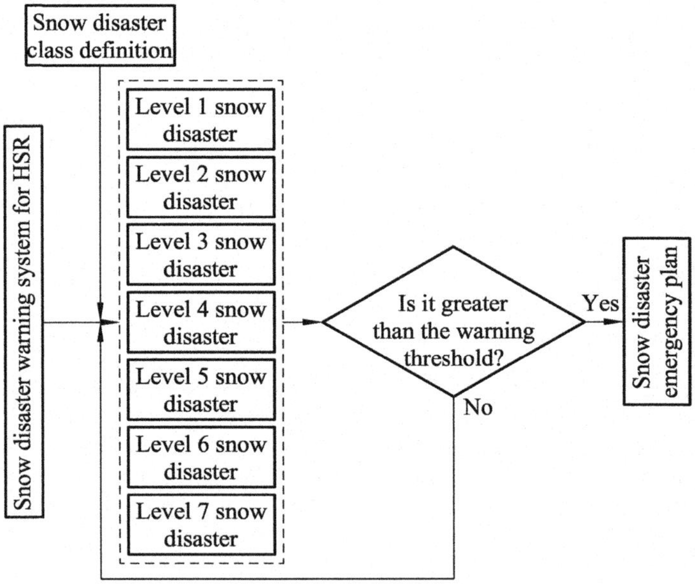

Fig. 8.4 HSR snow disaster warning system

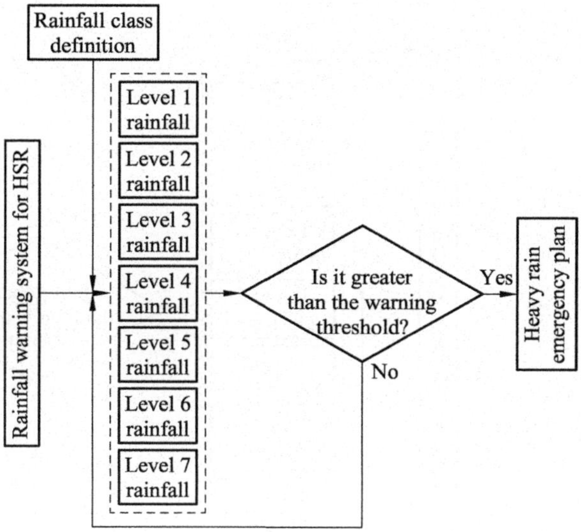

Fig. 8.5 HSR rainstorm warning system

(4) **HSR geological warning system**. In the HSR safety operation, in order to stop geological disasters from harming the HSR safety operation, a variety of cutting-edge technologies and efficient methodologies are employed to assess and monitor the dynamic changes of geological disaster activities and other triggering elements. To effectively monitor the HSR safety operation, the geological monitoring system of the HSR primarily records the change process of various precursor phenomena prior to the occurrence of geological disasters and the activity process after the occurrence of geological disasters through instrument measurement. We can understand the evolution characteristics of geological disasters, timely discover the microchanges such as slope ground cracking, peeling, ground swelling, sudden turbidity of spring water, changes in flow, skew of trees and wall cracking, timely capture the precursor information of geological disasters and give the early warning of geological disasters by regularly monitoring whether there is any abnormal change in the geological disaster potential sites. This chapter constructs the HSR geological warning system according to the research results at home and abroad, as shown in Fig. 8.6.

(5) **HSR earthquake warning system**. In the HSR safety operation, earthquakes damage high-speed lines, bridges and tunnels and other infrastructure, making it possible for HSR accidents and deaths to occur. High-speed trains must be halted when the peak acceleration of local vibration is equal to or more than 0.1 g, per the pertinent research report of the Beijing–Shanghai HSR. Therefore, to a certain extent, it is necessary to take real-time monitoring measures for key sections of HSR operation lines. This chapter constructs the HSR earthquake warning system in accordance with the findings at home and abroad, as shown in Fig. 8.7.

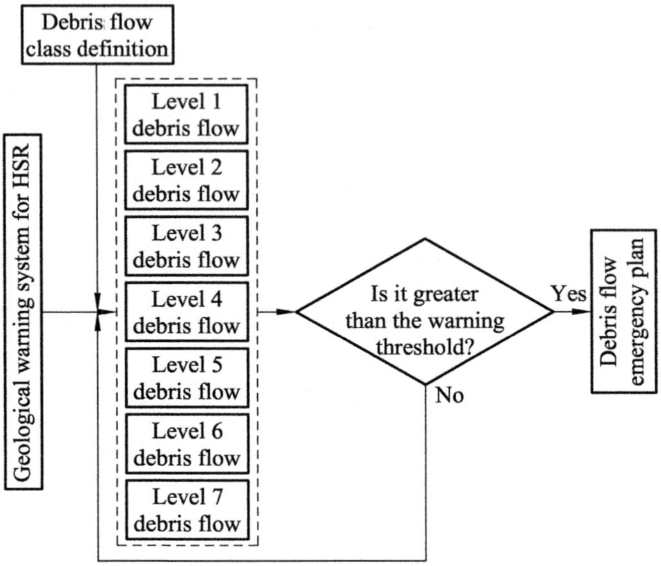

Fig. 8.6 HSR geological early warning system

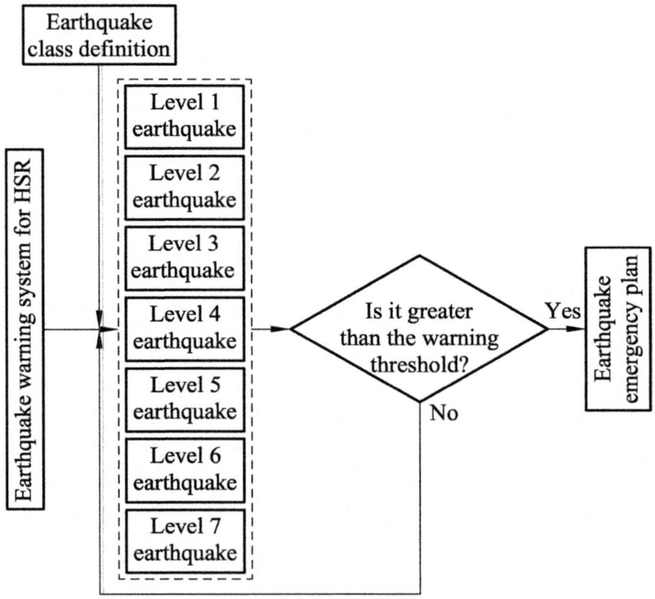

Fig. 8.7 HSR earthquake early warning system

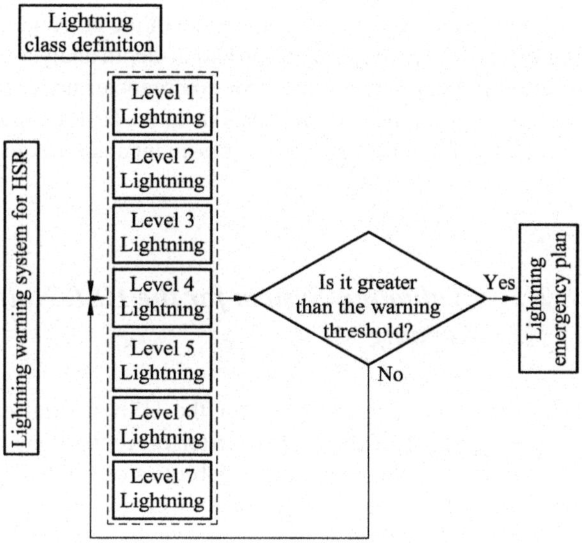

Fig. 8.8 HSR lightning warning system

(6) **HSR lightning warning system.** The HSR lightning monitoring system consists of a central station and several online TDOA detection stations across the HSR safety operation. When the lightning cloud discharges to the ground in the monitored area, the lightning monitoring central station of the HSR can calculate and determine the location of lightning strike point through special programs according to the time difference of lightning discharge electromagnetic signal obtained by each TDOA detection station. After a period of accumulation, it is possible to determine the frequency and density of ground lightning in the HSR region under observation, as well as the timing, position, amplitude and polarity of each lightning strike. This chapter constructs the HSR lightning warning system according to the research results at home and abroad, as shown in Fig. 8.8.

8.2 Natural Disaster Warning Model for the HSR Safety Operation

Natural disasters' effects on the HSR safety operation fall under the category of emergencies in this context. In order to fulfill the timeliness needs of the emergency decision-making and to assure the HSR safety operation, the quick comprehensive evaluation of emergencies is helpful in the early stages of natural disaster situations. In order to address the problem that the data volume of the HSR emergencies under natural disasters is large, the data update is fast, and the calculation process is too complex to meet the timeliness requirements of emergency decision-making, this chapter constructs a rapid assessment model for the HSR natural disaster early warning. The rapid assessment model of the HSR natural disaster warning solves the problem of rapid discrimination and in-depth analysis of emergency-related information at the early stage of the HSR emergencies under natural disasters, especially the information acquisition and analysis of the HSR emergencies under natural disasters. It handles the multifactor risk assessment and emergency decision-making dilemma under difficult situations.

8.2.1 Evaluation System for the HSR Natural Disaster Warning

I. Rapid assessment index for the HSR natural disaster warning. An essential component of the HSR disaster preventive and mitigation management is the emergency management capabilities of HSR situations during natural catastrophes. This chapter develops a three-layer rapid assessment index system for the HSR natural disaster warning, as shown in Fig. 8.9. The first layer is the general objective layer; the second layer contains six subsystems (crosswind system, earthquake system, temperature system, geological system, rainfall system and lightning system) for measuring the emergency response capability of the HSR under natural disasters; the third layer is the evaluation index system and decision-making model. The following are the guidelines for developing a rapid assessment system for the HSR natural disaster warning.

(1) Simple principle of the early warning system: the HSR early warning system for emergencies under natural disasters is oriented to the users, who require a

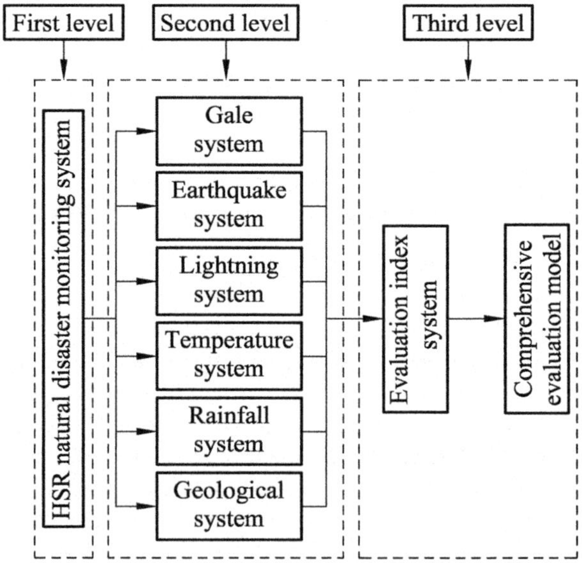

Fig. 8.9 Comprehensive HSR early warning system for natural disasters

straightforward human–machine interface. For dispatchers to make quick judgments and choices, the information supplied by the early warning system of HSR situations during natural disasters should be straightforward, quick, and accurate.

(2) Reliability principle of the early warning system: Because each alarm message provided by the early warning system of HSR crises under natural disasters is strongly tied to traffic safety and efficiency, the early warning system of HSR emergencies under natural disasters must be highly reliable.

(3) Integration principle of the early warning system: There are many types of early warning subsystems for HSR emergencies under natural disasters, and the on-site facilities are scattered. All subsystems should be integrated as much as feasible and, where circumstances allow, they should share resources with other system equipment in order to save costs and simplify maintenance and management.

According to Fig. 8.9, in the HSR safety operation, this chapter constructs a rapid assessment index system for HSR natural disaster warning, as shown in Fig. 8.10.

II. **Rapid assessment model of natural disasters for the HSR safety operation.** In the early stages of the HSR's natural disaster events, we are faced with the uncertainty of the information about the disasters, the variability of the disaster data, the scarcity of resources, other practical challenges, as well as the enormous risk of secondary events and derivative disaster events. In order to effectively respond to catastrophes, it is crucial to accurately understand the HSR operating status through the quick and in-the-moment comprehensive evaluation model of the HSR natural disaster early warning.

In the HSR safety operation, the comprehensive and rapid assessment model of the HSR natural disaster early warning should have the characteristics of being

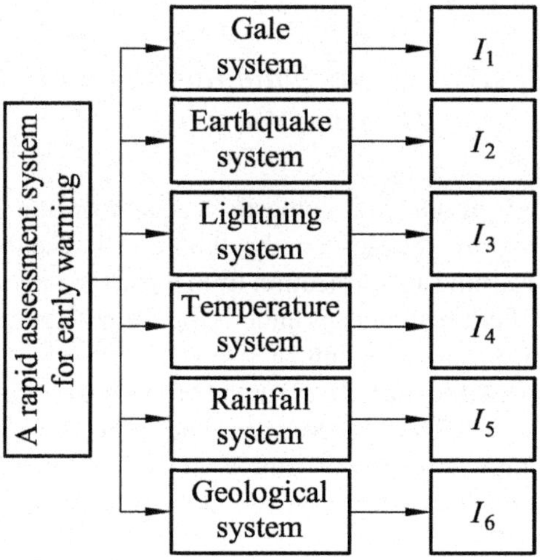

Fig. 8.10 Evaluation index system of the HSR natural disaster early warning

free from the limitation of data size, simple and easy to operate, so as to meet the requirements of emergency decision-making on the assessment speed in the early stage of natural disasters, when the data are dynamically updated and the data size is increasing. Three basic steps, including the data input stage, the data processing stage, and the data output stage, comprise the rapid assessment process for the HSR natural catastrophe warning (Fig. 8.9).

Stage 1: Data input stage. In the data input stage of the HSR natural disaster early warning, the HSR emergency data under natural disasters are dynamically collected and standardized to generate a dynamic data set to prepare for the next step of the data analysis;

Stage 2: Data processing stage. Two factors—how to increase the effectiveness of the design of the evaluation index system and how to implement the dynamic and real-time data update—reflect the timeliness of the data processing stage of the HSR natural disaster early warning. This phase utilizes an integrated data processing paradigm, which is the most crucial phase of the rapid assessment process for HSR emergencies during natural disasters;

Stage 3: Data output stage. In the data output stage of the HSR natural disaster early warning, according to the comprehensive evaluation model of the HSR natural disaster early warning, the evaluation results are given and decision-making suggestions are put forward.

In short, the comprehensive and rapid assessment model of the HSR natural disaster warning can address the problem that when the size of the HSR emergencies under natural disasters data is large, the calculation process is too complex and challenging to meet the high timeliness requirements of emergency decision-making at the beginning of the emergency. It enables the new model to accomplish speedy comprehensive emergency assessment in terms of process simplification and simple operation in emergency decision-making.

8.2.2 *An Attribute Identification Model for the HSR Natural Disaster Warning*

At present, the domestic research on the impact of the HSR natural environment mainly focuses on the theoretical research of natural disaster emergency early warning and protection from the macro-perspective, while the comprehensive studies of the mechanisms underlying natural environmental factors from a microperspective are scarce. In contrast, overseas research on the impact of the natural environment on the HSR started earlier and has already established an effective early warning system for HSR natural disasters. Therefore, this chapter conducts early warning research on natural disasters for the HSR safety operation in China using the attribute identification theory. It primarily conducts the comprehensive analysis and judgment in various environments, providing a theoretical basis for the safe operation and protection of the HSR in China.

I. **The attribute identification theory for the HSR natural disaster warning.** Professor Cheng Qiansheng of Peking University introduced attribute mathematics, which employs attribute connections to explain the general movement law of objects and substances as well as the associated structural relations in the process of change. Attribute mathematics is a mathematical science that expresses the overall movement attribute relationship between things and substances, the law and connotation of mutual change, and the analysis of development trend based on human research on the overall movement, continuous change, and infinite development law of natural numbers.

This chapter takes 31 provinces, municipalities, and autonomous regions in China as the sample space $X = \{x_1, x_2, x_3, \cdots, x_{31}\}$ for calculation, and the six rapid assessment indicators $\{I_1, I_2, I_3, I_4, I_5, I_6\}$ of the HSR natural disaster warning in Fig. 8.10 are taken as the assessment indicator I_j of the HSR natural disaster impact in each region.

The value of the jth natural disaster impact assessment indicator of the ith region is expressed as $x_{ij} = (i = 1, 2, \cdots, 31;\ j = 1, 2, 3, 4, 5, 6)$.

F is defined as an ordered segmentation set on the sample space X.

The impact of natural disasters is divided into four categories: especially serious, more serious, serious and moderate. Then the ordered segmentation set is $F = \{C_1, C_2, C_3, C_4, C_5\}$, where $C_1 > C_2 > C_3 > C_4 > C_5$ is satisfied.

The ordered segmentation set is the set of thresholds of each natural disaster evaluation index for the segmentation class, so the ordered segmentation set can be expressed in the following standard forms according to the definition:

$$
\begin{array}{c|ccccc|}
 & C_1 & C_2 & C_3 & C_4 & C_5 \\
I_1 & a_{11} & a_{12} & a_{13} & a_{14} & a_{15} \\
I_2 & a_{21} & a_{22} & a_{23} & a_{24} & a_{25} \\
\vdots & \vdots & \vdots & \vdots & \vdots & \vdots \\
I_6 & a_{61} & a_{62} & a_{63} & a_{64} & a_{65}
\end{array},
$$

where $a_{ij} = (i = 1, 2, \cdots, 31;\ j = 1, 2, 3, 4, 5, 6)$ represents the threshold value of the evaluation index I_i of the HSR natural disaster early warning in the attribute class of C_j, meeting $a_{i1} > a_{i2} > a_{i3} > a_{i4} > a_{i5}$.

II. **The assessment value of standard attributes of the HSR natural disaster warning.** In the HSR safety operation, the assessment value of standard attributes of the HSR natural disaster early warning is the quantity representing the attribute degree of the jth index x_{ij} of a high-speed railway X_i, expressed as $u_{ijk} = u(x_{ijk} \in C_k)$. One of them is the choice of attribute measure function, which is the primary component of attribute recognition and directly connected to the outcome of attribute discrimination. The essence of the attribute evaluation is to determine the multi-dimensional space distance between indicators and attribute classes. The calculation steps of attribute evaluation values are as follows:

Step 1. Calculate the Mahalanobis distance between the HSR and attribute class. Assuming that a regional HSR's assessment index for its natural disaster factor has been determined, the Mahalanobis distance between the regional HSR X_i and the attribute class C_k is

$$d_{ik} = \sqrt{(X_i - C_k) \sum\nolimits_{ik} (X_i - C_k)^T}, \tag{8.1}$$

where

$X_i = (x_{i1}, x_{i2}, \cdots, x_{i6})$	Is the vector of natural hazard factor evaluation indicators of the ith region;
$C_k = (a_{k1}, a_{k2}, \cdots, a_{k6})$	Is the classification standard vector of each natural disaster factor evaluation index in the attribute class k.
Σ_{ik}	Represents the covariance matrix of X_i and C_k:

$$\Sigma_{ik} = \begin{bmatrix} \mathrm{Cov}(x_{i1}, a_{k1}) & \mathrm{Cov}(x_{i1}, a_{k2}) & \cdots & \mathrm{Cov}(x_{i1}, a_{k6}) \\ \mathrm{Cov}(x_{i2}, a_{k1}) & \mathrm{Cov}(x_{i2}, a_{k2}) & \cdots & \mathrm{Cov}(x_{i2}, a_{k6}) \\ \vdots & \vdots & \ddots & \vdots \\ \mathrm{Cov}(x_{i6}, a_{k1}) & \mathrm{Cov}(x_{i6}, a_{k2}) & \cdots & \mathrm{Cov}(x_{i6}, a_{k6}) \end{bmatrix} \tag{8.2}$$

Step 2. Calculate the standard attribute assessment value of regional HSR. In the HSR safety operation, it is assumed that the Mahalanobis distance d_{ik} of an HSR X_i in each attribute class C_k has been calculated. In general, the greater the Mahalanobis distance, the lower the resemblance between the HSR and the attribute class, and the lower the assessment value. Therefore, if the reciprocal of the Mahalanobis distance $\frac{1}{d_{ik}}$ is taken as the evaluation value of samples and attribute categories, the standard attribute evaluation value of high-speed railway evaluation X_i in the evaluation class C_k is:

$$u_{ik} = \frac{1}{d_{ik}} \left[\sum_{j=1}^{5} \frac{1}{d_{ij}} \right]^{-1}. \tag{8.3}$$

III. **The safety of the HSR natural disaster warning.** The evaluation value of standard attributes of a regional HSR for a certain evaluation class can be obtained through Eq. (8.3), according to which the attribute recognition and score calculation of the regional HSR can be carried out. In the attribute recognition, the evaluation class is carried out according to the confidence criterion; that is, an interval value of λ is set. If:

$$k_i = \min\{k : \sum_{l=1}^{k} u_{il} \geq \lambda, k = 5, 4, 3, 2, 1\}. \quad (8.4)$$

It can be considered that the HSR X_i belongs to the evaluation class C_k, which is $0.6 \leq \lambda \leq 0.7$ in the medium case. Assuming that the score corresponding to each evaluation class C_k is q_k, the comprehensive safety value of the HSR X_i is:

$$S_i = \sum_{k=1}^{4} u_{ik} q_k. \quad (8.5)$$

Equation (8.5) obtains the safety score of the natural disaster factor of the HSR. Assuming that $S_i > S_j$, the HSR natural disaster factor X_i is higher than the safety level of the HSR X_j. Therefore, the impact of natural disaster factors in different regions on the safety of the HSR can be determined according to the safety score of the HSR natural disaster early warning.

8.2.3 Rough Set Prediction Model for the HSR Natural Disaster Warning

The unproven mathematics idea of rough set is utilized for quantitative study with the goal of early warning of natural disasters in the HSR safety operation. The prediction index system of the HSR natural disaster early warning is built on the basis of numerous variables of the HSR safety operation. To develop the rough prediction model of the HSR natural catastrophe early warning, the rough set and unascertained prediction are organically blended by integrating qualitative and quantitative analyses. Finally, the prediction matrix is produced to convert the HSR system's multi-level measurement into a single-level measurement on the basis of the weight coefficient of the prediction index, and the predictions for the HSR natural disaster early warning are given as numerical values. The prediction values of the HSR natural disaster early warning can concisely and accurately reflect the current level of the HSR safety operation.

The prediction index system of the HSR natural disaster early warning is an essential basis for characterizing and estimating the amount of natural disasters incurred by the HSR in the HSR safety operation. Therefore, on the basis of building the prediction index system of the HSR natural disaster early warning, the rough set theory is applied to simplify the prediction indexes of the HSR natural disaster early warning and extract the main prediction indexes affecting the HSR safety operation, and then the prediction model is used for comprehensive prediction. The entire prediction results of the HSR natural disaster early warning are achieved in accordance with the single-index prediction matrix and assessment criteria based on the weight coefficient of the prediction index determined by the objective weighting approach.

I. **The rough set prediction theory of the HSR natural disaster warning.** Let $x_1, x_2, \cdots, x_n$ represent n high-speed railways to be evaluated for safe operation, which is denoted by $X = \{x_1, x_2, \cdots, x_n\}$, and is called the domain of the HSR natural disaster early warning. The six-index system $I_1, I_2, \cdots, I_6$ (Fig. 8.10) for predicting the HSR $x_i (x_i \in X)$ is recorded as $I = \{I_1, I_2, \cdots, I_6\}$.

x_{ij} represents the observation value of the HSR x_i under the prediction index I_j of the HSR natural disaster early warning.

$C = \{c_1, c_2, \cdots, c_k\}$ is set as the evaluation space for the HSR natural disaster early warning, where $c_k (1 \le k \le K)$ is the kth evaluation level of the HSR natural disaster early warning.

Therefore, a prediction index system for the early warning of HSR natural disasters is built in accordance with the elements impacting the HSR safety operation. A thorough forecast of the HSR safety operation is created using the rough set comprehensive assessment technique based on pertinent national standards and pertinent statistical data, and the prediction results are given by the quantitative value. The prediction results of the HSR natural disaster early warning can closely link the qualitative and quantitative information, historical accident data and assessment status information of the railway safety operation, and dynamically form an organic prediction system.

II. **Single-indicator prediction of the HSR natural disaster warning.** In the HSR safety operation, if the observation value x_{ij} of the prediction index I_j of the HSR x_i on the HSR natural disaster early warning is different, the prediction index of the HSR natural disaster early warning makes the degree of the HSR x_i at each evaluation level different.

Let $\mu_{ijk} = \mu\left(x_{ij} \in c_k\right)$ be the degree to which x_{ij} makes the HSR x_i in the kth assessment level c_k of the operation safety assessment, then μ_{ijk} meets the following requirements:

$$0 \le \mu_{ijk} \le 1, \text{and } \mu\left[x_{ij} \in \bigcup_{k=1}^{K} c_k\right] = \sum_{k=1}^{K} \mu(x_{ij} \in c_k), \mu(x_{ij} \in c) = 1$$

where $i = 1, 2, \cdots, n; j = 1, 2, \cdots, 6; k = 1, 2, \cdots, K$.

μ_{ijk} is called the unknown prediction of the HSR natural disaster warning, referred to as the prediction of the HSR natural disaster warning, that is

$$(\mu_{ijk})_{m\times K} = \begin{bmatrix} \mu_{i11} & \mu_{i12} & \cdots & \mu_{i1K} \\ \mu_{i21} & \mu_{i22} & \cdots & \mu_{i2K} \\ \vdots & \vdots & \cdots & \vdots \\ \mu_{im1} & \mu_{im2} & \cdots & \mu_{imK} \end{bmatrix}, (i = 1, 2, \cdots, n) \tag{8.6}$$

$(\mu_{ijk})_{m\times K}$ is the single-indicator prediction matrix of the HSR, where $\mu^i_j (1 \le j \le 6)$ denotes the unknown prediction that x_{ij} makes x_i at each evaluation level.

III. **Weight coefficients of the HSR natural disaster warning.** Due to the complexity of the HSR system, it is not feasible to obtain a large number of training samples, use the objective weighting method, or compute the correlation coefficients of the attribute prediction vectors to determine the weight values of the HSR natural disaster warning prediction indexes. In order to increase the objectivity of the decision-making process, this chapter adopts the fuzzy dispersion weighting method to determine the weight coefficients of the prediction indexes of the HSR natural disaster warning. Then the weight coefficients of the prediction index I_i of the HSR natural disaster warning are:

$$w_i = \Delta_i \cdot \left[\sum_{i=1}^{6} \Delta_i\right]^{-1}, \tag{8.7}$$

where $\Delta_i = \frac{1}{n}\sum_{j=1}^{n} \left|b_{ji} - \overline{b_i}\right|$, and $\overline{b_i} = \frac{1}{n}\sum_{j=1}^{n} b_{ji}$.

IV. **Comprehensive prediction of the HSR natural disaster warning.** From Eq. (8.7), we can know the single-index prediction matrix of the safety operation of the HSR x_i, and the weight coefficients of each prediction index of the HSR x_i, so that we can let

$$\begin{aligned} \mu^i &= W \cdot (\mu_{ijk})_{m\times K} \\ &= \left(w_1, w_2, \cdots, w_m\right) \cdot \begin{bmatrix} \mu_{i11} & \mu_{i12} & \cdots & \mu_{i1K} \\ \mu_{i21} & \mu_{i22} & \cdots & \mu_{i2K} \\ \vdots & \vdots & \cdots & \vdots \\ \mu_{im1} & \mu_{im2} & \cdots & \mu_{imK} \end{bmatrix} \\ &= (\mu_{i1}, \mu_{i2}, \cdots, \mu_{iK}) \end{aligned} \tag{8.8}$$

Then μ^i is the vector of safety operation prediction for the HSR.

V. **Prediction criterion of the HSR natural disaster warning.** The evaluation level of the HSR natural disaster warning is ordered. The kth evaluation level c_k of the prediction is "better than" the k + 1th evaluation level c_{k+1} of the prediction. Therefore, the maximum prediction recognition criterion is inappropriate and should be replaced with the confidence recognition criterion.

Criterion 1 of the HSR natural disaster warning: confidence identification criterion.

Confidence identification criterion: The confidence level of the HSR natural disaster warning is λ, which is usually taken as 0.6, so that

$$k_o = \min_k \left[\left(\sum_{l=1}^{k} \mu_{il} \right) \geq \lambda, k = 1, 2, \cdots, K \right]. \tag{8.9}$$

Then, the HSR x_i is determined to belong to the k_o th assessment level c_{k_o} in the safety operation.

Criteria 2 of the HSR natural disaster warning: ranking value criterion.

$$p_i = \max(\mu_{i1}, \mu_{i2}, \cdots, \mu_{iK}) \tag{8.10}$$

According to the prediction model of the HSR natural disaster warning, the comprehensive prediction value p_i of the safety operation of the ith HSR x_i can be obtained. According to the prediction theory, the larger the prediction value p_i of the HSR natural disaster early warning, the better the HSR safety operation, and the less the impact of natural disasters. On the contrary, the smaller the predictive value p_i of the HSR natural disaster early warning, the worse the HSR safety operation and the greater the impact of natural disasters. Therefore, on the basis of judging the safe operation situation of each HSR, according to the prediction value p_i of the HSR natural disaster warning, the safe operation situation of m high-speed railways can also be ranked. The higher the ranking, the safer the safe operation situation of the high-speed railway is, and the less natural disasters it is susceptible to.

8.2.4 Complex Matter-Element Model for the HSR Natural Disaster Warning

The quantitative values of each natural disaster component in the HSR safety operation are provided in accordance with the theory of matter-element analysis, based on the pertinent national standards and pertinent statistical data. The complex element analysis model of the HSR natural disaster warning is established, and the measurement results of the HSR safety operation are expressed in quantitative values. The complex element concept is used to establish the prediction model of the HSR natural disaster warning. Not only can the model organically link qualitative and quantitative information, historical accident data, and management status in HSR safety management information, but it can also fully reflect such a scientific prediction mechanism as focusing on norms, emphasizing the status quo and following facts in the measurement process, and objectively reflect the safety status quo of the HSR.

(1) **Complex matter-element theory of the HSR natural disaster warning.** According to the extension theory proposed by Professor Cai Wen, matter-element refers to the basic element of a thing described by an ordered triple: "thing, feature, value." Therefore, given the name N of a thing, its characteristic C and its magnitude V can form a basic matter-element $R = (N, C, V)$.

If the thing N has n characteristics $c_1, c_2, \cdots, c_n$ and corresponding quantity $v_1, v_2, \cdots, v_n$, $R = (N, C, V)$ is called n-dimensional matter-element. In addition, if things in matter-elements are schemes and characterized by the information, they are called complex matter-elements and recorded as $\underset{\sim H}{R}$. If there are m schemes in the complex matter-element, it is the complex matter-element of m schemes, recorded as $\underset{\sim mH}{R}$.

(2) **Prediction model of the HSR natural disaster warning.** There are m high-speed operating railways, which are recorded as $R_1, R_2, \cdots, R_m$. The HSR natural disaster early warning prediction indicators (six evaluation indicators in Fig. 8.10 are used as prediction indicators) are used as characteristics, and their corresponding prediction values are used to form a complex matter-element of the HSR natural disaster early warning. Then the complex matter-element of natural disaster early warning for m high-speed railways is:

$$\underset{\text{HSR}}{HR} = \begin{bmatrix} & R_1 & R_2 & \cdots & R_m \\ C_1 & x_{11} & x_{21} & \cdots & x_{m1} \\ C_2 & x_{12} & x_{22} & \cdots & x_{m2} \\ \vdots & \vdots & \vdots & \vdots & \vdots \\ C_6 & x_{1,6} & x_{2,6} & \cdots & x_{m,6} \end{bmatrix}, \quad (8.11)$$

where

R_i is the ith high-speed operating railway;
C_j is the ith prediction index of the HSR natural disaster warning, and its corresponding quantity value is denoted by x_{ij}.

The HSR natural disaster early warning model based on the complex matter-element is calculated as follows:

Step 1: Standardize the value of the complex matter-element in the HSR natural disaster early warning. Due to the different meanings of each prediction index in the HSR natural disaster early warning and the different calculation methods of the prediction index value in Eq. (8.11), the HSR natural disaster warning has various dimensions for each prediction index. Therefore, in order to ensure the commensurability of each prediction index in the HSR natural disaster early warning, its value must be standardized.

Let $J^+ = \{\text{benefit predictor}\}$, $J^- = \{\text{cost predictor}\}$. Then

$$\mu_{ij} = \left(x_{ij} - \min_{1 \le i \le n} x_{ij}\right) \Big/ \left(\max_{1 \le i \le n} x_{ij} - \min_{1 \le i \le n} x_{ij}\right), (i = 1, 2, \ldots, n;\ j \in J^{+}) \tag{8.12}$$

$$\mu_{ij} = \left(\max_{1 \le i \le n} x_{ij} - x_{ij}\right) \Big/ \left(\max_{1 \le i \le n} x_{ij} - \min_{1 \le i \le n} x_{ij}\right), (i = 1, 2, \ldots, n;\ j \in J^{-}). \tag{8.13}$$

After standardization, the complex matter-element of m high-speed operating railways is

$$\underset{\sim \mathrm{HSR}}{HR} = \begin{bmatrix} & R_1 & R_2 & \cdots & R_m \\ C_1 & \mu_{11} & \mu_{21} & \cdots & \mu_{m1} \\ C_2 & \mu_{12} & \mu_{22} & \cdots & \mu_{m2} \\ \vdots & \vdots & \vdots & \vdots & \vdots \\ C_6 & \mu_{1,6} & \mu_{2,6} & \cdots & \mu_{m,6} \end{bmatrix}. \tag{8.14}$$

Step 2: Determine the weight coefficients of the prediction indicators in the HSR natural disaster warning. Due to the complexity of the HSR systems and the uncertainty of natural disasters, it is often impossible to clearly give the weight information of each prediction index in the HSR natural disaster early warning. It is frequently unable to provide the weight information of each prediction index in the HSR natural disaster early warning explicitly due to the complexity of the HSR systems and the unpredictability of natural disasters. The details are as follows:

When $y_j = \max_{1 \le i \le m} \mu_{ij}, j = 1, 2, \cdots, 6; i = 1, 2, \cdots, m$, there is a reference sequence $Y = \{y_1, y_2, \cdots, y_6\}$. The correlation coefficient of the prediction index C_j in item j of the complex matter-element $\underset{\sim \mathrm{HSR}}{HR}$ for the HSR natural disaster warning is:

$$\zeta_{ij} = \frac{\min_i \min_j \left|\mu_{ij} - y_j\right| + 0.5 \max_i \max_j \left|\mu_{ij} - y_j\right|}{\left|\mu_{ij} - y_j\right| + 0.5 \max_i \max_j \left|\mu_{ij} - y_j\right|}. \tag{8.15}$$

Therefore, according to the information theory, the entropy of the jth prediction index C_j in the HSR natural disaster early warning is:

$$F_j = -\frac{1}{\ln 6} \sum\nolimits_{i=1}^{m} \frac{\zeta_{ij}}{\sum_{i=1}^{m} \zeta_{ij}} \ln \frac{\zeta_{ij}}{\sum_{i=1}^{m} \zeta_{ij}}. \tag{8.16}$$

Since $F_j \in [0, 1]$, if the deviation degree is $k_j = 1 - F_j$, then the weight coefficient of the jth prediction index C_j in the HSR natural disaster early warning is:

$$w_j = \frac{k_j}{\sum_{j=1}^{n} k_j}. \tag{8.17}$$

Due to the weight coefficient w_j of the jth prediction index C_j in the HSR natural disaster early warning, the complex matter-element of the prediction index weight in the HSR natural disaster early warning is constructed.

$$\underset{wj}{HR} = \begin{bmatrix} & C_1 & C_2 & \cdots & C_6 \\ w_j & w_1 & w_2 & \cdots & w_6 \end{bmatrix} \tag{8.18}$$

Step 3: Determine the complex correlated entropy element of the HSR natural disaster warning system. In the HSR safe operation, the complex matter-element of the HSR natural disaster warning $\underset{\sim HSR}{HR}$ system can be determined by Eqs. (8.14) and (8.15), namely

$$\begin{aligned} \underset{\sim HSR}{HR} &= \begin{bmatrix} & R_1 & R_2 & \cdots & R_i & \cdots & R_m \\ H_i & H_1 & H_2 & \cdots & H_i & \cdots & H_m \end{bmatrix} \\ &= \left[\begin{matrix} R_1 & \cdots \\ H_i - \sum_{j=1}^{6} P(w_j\mu_{1j}) \ln P(w_j\mu_{1j}) & \cdots \end{matrix} \right. \\ &\quad \left. \begin{matrix} R_i & \cdots & R_m \\ - - \sum_{j=1}^{6} P(w_j\mu_{ij}) \ln P(w_j\mu_{ij}) & \cdots & - \sum_{j=1}^{6} P(w_j\mu_{mj}) \ln P(w_j\mu_{mj}) \end{matrix} \right] \end{aligned} \tag{8.19}$$

where $P(w_j\mu_{ij}) = w_i\mu_{ij} \cdot \left[\sum_{j=1}^{6} w_j\mu_{ij} \right]^{-1}$, $(i = 1, 2, \cdots, m; j = 1, 2, \cdots, 6)$.

Step 4: Determine the prediction value of the HSR natural disaster warning system. According to the complex matter-element model of the HSR natural disaster early warning system, the comprehensive prediction value $H_i (i = 1, 2, \cdots, m)$ of the HSR natural disaster early warning of the ith high-speed railway can be obtained.

The functional relationship of the HSR natural disaster warning is expressed as

$$H_i = \max_{1 \le i \le m} \{H_i\}. \tag{8.20}$$

According to the theory of matter-element analysis, in the HSR safe operation, the larger the integrated prediction value of the HSR natural disaster warning H_i system, the better the HSR safe operation situation under natural disasters; the smaller the integrated prediction value of the HSR natural disaster warning H_i system, the worse the HSR safe operation situation under natural disasters, which is very prone to danger and should be warned and prevented.

8.3 Integrated Natural Disaster Warning Mechanism for the HSR Safe Operation

In the HSR safe operation, although the probability of natural disasters such as crosswind, rainstorm, snow, earthquake, geological hazard and lightning is small, they can bring great harm. Based on the analysis of the current foreign HSR disaster prevention and safety warning system, we study the HSR disaster monitoring technology and development trend, especially hazards of crosswind, heavy rain, heavy snow, earthquake, geological disaster, lightning and other disasters to the HSR safe operation. In order to prevent or mitigate hazards of natural disasters and emergencies to the HSR safe operation, combining with the actual HSR situation in China, the overall structure of the HSR early warning system for natural disasters is proposed to realize the distributed acquisition, centralized management, emergency response, comprehensive analysis and application of real-time monitoring information of disasters such as crosswind, heavy rain, heavy snow, earthquake, geological disaster and lightning. The warning system can grasp the dynamics of natural disasters in time to ensure the HSR safe operation.

8.3.1 Integrated Functions of the Natural Disaster Warning

The HSR early warning system for natural disasters ensures the HSR safe operation. It monitors natural disasters (wind, rain, snow, earthquake, geological hazard, lightning, etc.) that endanger the safety of train operation in real time, collects and summarizes monitoring information of various monitoring equipment, achieves distributed acquisition, centralized management, and comprehensive application of the HSR monitoring information, fully grasps the dynamics of natural disasters and provides timely and accurate disaster alarm and early warning functions. It is an indispensable and important technical guarantee in the HSR transportation system to immediately take corresponding emergency response measures according to the severity of disasters, including preventing or reducing losses caused by natural disasters, avoiding secondary disasters, and providing data basis for HSR operation plan adjustment, traffic control, emergency rescue, maintenance and other work.

The main functions of the HSR early warning system under natural disasters include HSR monitoring network layout, alarm threshold setting, emergency response measures, monitoring equipment selection and application and emergency plan management. The system also provides the sharing and exchange of relevant basic data and monitoring data so as to master the status of natural disaster monitoring alarm and equipment operation, supervise and guide the operation of the disaster prevention and early warning system of each high-speed railway line. The decision support services are provided for the construction of the HSR disaster prevention and early warning system by analyzing the overall HSR disaster monitoring data.

8.3.2 System Architecture for the Natural Disaster Warning

The HSR early warning system under natural disasters consists of four parts: on-site monitoring points along the line (wind, rain, snow, earthquake, geological disaster and lightning monitoring equipment), early warning units, early warning centers and relevant system interfaces. It provides real-time monitoring, early warning and alarm functions for the HSR natural disasters and emergencies, so as to realize emergency response to natural disaster alarms, minimize losses caused by disasters and prevent secondary disasters.,

(1) **HSR real-time warning subsystem** It realizes real-time warning of natural disasters, such as wind, rain, snow, earthquake, geological disaster and lightning, monitoring equipment status, providing alarm and warning functions and recording processing results.
(2) **HSR statistical analysis subsystem** The effective information is generated through the statistical analysis of the monitored natural disaster data, alarm information and equipment status data of the HSR.
(3) **HSR equipment management subsystem** It realizes the management of the equipment of HSR early warning system for natural disasters, such as the wind speed and wind direction indicator, rain gauge, snow gauge, strong motion monitoring instrument, shock sensing cabinet, double cable sensor, early warning unit, server and switch, by mastering the operation status of equipment and realizes the maintenance management.
(4) **HSR emergency response subsystem** It provides alarm thresholds and alarm levels for the natural disasters such as wind, rain, snow, earthquake, geological disaster and lightning and develops corresponding emergency response measures for various HSR alarm levels.
(5) **HSR system management and maintenance subsystem** It provides support for the operation of the HSR natural disaster warning system, including the receipt and verification of the detected data, information exchange with other systems, system parameter configuration and safety management, system operation warning, data dumping and backup.

In the HSR safe operation, the HSR early warning system under natural disasters is a new field of the railway informatization. It is closely related to the HSR operation safety and has the characteristics of strong real time, high reliability, wide coverage, high professionalism and long optimization period. Therefore, the establishment of the HSR early warning system under natural disasters meets the requirements of the

HSR in China, with a significant social and economic significance to ensure the HSR safe operation in China.

8.4 Natural Disaster Warning System for the HSR Safe Operation

The HSR natural disaster early warning system is one of the important basic equipments to ensure the HSR safe and high-speed operation. The system is built on the communication and transmission system. It monitors and detects disasters endangering railway transportation through the on-site monitoring equipment, such as wind, rain, lightning, temperature, geological disaster and earthquake. It also provides the processed disaster warning information, speed limit information or shutdown information as the basis for the decision-making of the dispatching center to ensure the safe and efficient operation of trains promptly. Moreover, the HSR early warning system for natural disasters is mainly composed of on-site monitoring equipment, early warning equipment, early warning data processing equipment, dispatching station equipment and other equipment.

8.4.1 Subsystems for the Natural Disaster Warning System

The on-site monitoring equipment of the HSR early warning system for natural disasters is mainly used to collect data on wind speed and direction, rainfall, temperature, geological disaster, lightning and earthquake. Therefore, it consists of six subsystems including crosswind early warning system, rainfall early warning system, temperature early warning system, earthquake early warning system, geological disaster early warning system and lightning early warning system for natural disasters.

(1) **HSR crosswind early warning system** The main component of HSR crosswind early system is the ultrasonic aerovane, which is used to collect the local wind speed and direction data. The equipment is set up on the windward side of lines and is installed at a height of 4 ± 0.1 m from the rail surface on catenary poles with hoops. With strong anti-electromagnetic interference, waterproof and dustproof performance and other functions, it can be used in complex and harsh environments. For the HSR, the aerovane shall be set in areas where the annual average maximum instantaneous wind speed is no more than 30 m/s and no less than 15 m/s, as shown in Table 8.1.

Since the impact of crosswinds on high-speed trains requires manual measurement and the operation of trains is controlled through dispatching, the crosswind early warning system needs to have a function of forecasting.

Table 8.1 Wind speed parameters in cross wind early warning system

Level	Wind speed/(m/s)		Speed limit/(km/h)	
	Wind speed interval	Early warning threshold	Speed interval	Early warning threshold
Level 1	Wind speed < 15	–	300 < Speed ≤ 350	350
Level 2	15 ≤ Wind speed < 20	15	250 < Speed ≤ 300	300
Level 3	20 ≤ Wind speed < 23	20	200 < Speed ≤ 250	250
Level 4	23 ≤ Wind speed < 26	23	150 < Speed ≤ 200	200
Level 5	26 ≤ Wind speed < 27	26	100 < Speed ≤ 150	150
Level 6	27 ≤ Wind speed < 30	27	50 < Speed ≤ 100	100
Level 7	30 ≤ Wind speed	30	Stop running	50

(2) **HSR rainfall early warning system** The main component of HSR rainfall early warning system is rain gauge, which is used to collect local rainfall data, as shown in Table 8.2.

(3) **HSR earthquake warning system** The main component of HSR earthquake warning system is the strong seismograph. The equipment can send timely and accurate warning for earthquake waves. When a line encounters an earthquake with the intensity greater than 6 (ground motion acceleration > 0.04 g, equivalent to a 5-magnitude earthquake), the earthquake warning system can automatically send alarm information, and control high-speed trains in the earthquake area to slow down or stop running through the high-speed train control system, as shown in Table 8.3.

When the HSR earthquake early warning system detects that the intensity of ground motion reaches the earthquake alarm threshold, it will send an alarm message to the HSR dispatching center and immediately trip the main circuit breaker of the substation to cut off the power supply to the catenary and force high-speed trains to stop. At the same time, the earthquake early warning system continues to monitor the subsequent ground motion acceleration and provide the HSR dispatching center with the basis for the train operation control after stopping.

(4) **HSR temperature early warning system** HSR temperature early warning system mainly includes the high-temperature early warning system and low-temperature early warning system. The high-temperature early warning system mainly sends warning for temperature in the desert and other high heat areas (such as China's Xinjiang the desert area in Xinjiang, China, where the temperature in summer reaches more than 60 °C) to ensure the HSR safe operation, as shown in Table 8.4; the low-temperature early warning system mainly sends warnings for temperature in the severe cold and other low-temperature areas (such as the northeast area in China, where the temperature in winter is more than 50°C below zero) to ensure the HSR safe operation, as shown in Table 8.5.

Table 8.2 Wind speed parameters in the rainfall early warning system

Level	Threshold of rainfall/(mm)		Speed limit/(km/h)	
	Name	Early warning threshold	Speed interval	Early warning threshold
Level 1	Hourly rainfall	25	300 < Speed ≤ 350	350
	Daily rainfall	90		
	Continuous rainfall	110		
Level 2	Hourly rainfall	30	250 < Speed ≤ 300	300
	Daily rainfall	100		
	Continuous rainfall	120		
Level 3	Hourly rainfall	35	200 < Speed ≤ 250	250
	Daily rainfall	110		
	Continuous rainfall	130		
Level 4	Hourly rainfall	40	150 < Speed ≤ 200	200
	Daily rainfall	120		
	Continuous rainfall	140		
Level 5	Hourly rainfall	45	100 < Speed ≤ 150	160
	Daily rainfall	135		
	Continuous rainfall	160		
Level 6	Hourly rainfall	55	50 < Speed ≤ 100	100
	Daily rainfall	182		
	Continuous rainfall	180		
Level 7	Hourly rainfall	60	Stop running	50
	Daily rainfall	200		
	Continuous rainfall	200		

Table 8.3 Magnitude parameters in earthquake early warning system

Level	Acceleration of ground motion/g		Speed limit/(km/h)	
	Magnitude interval	Early warning threshold	Speed interval	Early warning threshold
Level 1	Acceleration ≤ 15	–	300 < Speed ≤ 350	350
Level 2	15 < Acceleration ≤ 20	15	250 < Speed ≤ 300	300
Level 3	20 < Acceleration ≤ 23	20	200 < Speed ≤ 250	250
Level 4	23 < Acceleration ≤ 26	23	150 < Speed ≤ 200	200
Level 5	26 < Acceleration ≤ 27	26	100 < Speed ≤ 150	150
Level 6	27 < Acceleration ≤ 30	27	50 < Speed ≤ 100	100
Level 7	30 < Acceleration	30	Stop running	50

Table 8.4 High-temperature parameters in the temperature early warning system

Level	Temperature/°C		Speed limit/(km/h)	
	Temperature interval	Early warning threshold	Speed interval	Early warning threshold
Level 1	Temperature $\leq$ 15	–	300 < Speed $\leq$ 350	350
Level 2	15 < Temperature $\leq$ 25	15	250 < Speed $\leq$ 300	300
Level 3	25 < Temperature $\leq$ 30	25	200 < Speed $\leq$ 250	250
Level 4	30 < Temperature $\leq$ 35	30	150 < Speed $\leq$ 200	200
Level 5	35 < Temperature $\leq$ 40	35	100 < Speed $\leq$ 150	150
Level 6	40 < Temperature $\leq$ 45	40	50 < Speed $\leq$ 100	100
Level 7	45 < Temperature	45	Stop running	50

Table 8.5 Low-temperature parameters in the temperature early warning system

Level	Temperature/°C		Speed limit/(km/h)	
	Temperature interval	Early warning threshold	Speed interval	Early warning threshold
Level 1	5 $\leq$ Temperature < 15	–	300 < Speed $\leq$ 350	350
Level 2	0 $\leq$ Temperature < 5	5	250 < Speed $\leq$ 300	300
Level 3	−10 $\leq$ Temperature < 0	0	200 < Speed $\leq$ 250	250
Level 4	−20 $\leq$ Temperature < -10	−10	150 < Speed $\leq$ 200	200
Level 5	−30 $\leq$ Temperature < -20	−20	100 < Speed $\leq$ 150	150
Level 6	−40 $\leq$ Temperature < -30	−30	50 < Speed $\leq$ 100	100
Level 7	Temperature < -40	−40	Stop running	50

(5) **HSR lightning early warning system** HSR lightning early warning system utilizes lightning positioning, radar, satellite, ground electric field, sounding and other observation data to identify, track, forecast and warn areas where lightning is likely to occur or has occurred through multi-source data fusion, statistical analysis, echo extrapolation-based method, numerical prediction and other technical means, as shown in Table 8.6.

(6) **HSR Geological Early Warning System** HSR geological early warning system records the change process of various precursors before the occurrence of geological disaster and the activity process after the occurrence of the geological disaster through direct observation and instrument measurement, as shown in Table 8.7.

Table 8.6 Lightning parameters in lightning early warning system

Level	Average annual number of days with lightning/day		Speed limit/(km/h)	
	Lightning interval	Early warning threshold	Speed interval	Early warning threshold
Level 1	Days with lightning < 10	–	300 < Speed ≤ 350	350
Level 2	10 ≤ Days with lightning < 20	10	250 < Speed ≤ 300	300
Level 3	20 ≤ Days with lightning~~Lightning~~ < 30	20	200 < Speed ≤ 250	250
Level 4	30 ≤ Days with lightning < 40	30	150 < Speed ≤ 200	200
Level 5	40 ≤ Days with lightning < 50	40	100 < Speed ≤ 150	150
Level 6	50 ≤ Days with lightning < 60	50	50 < Speed ≤ 100	100
Level 7	60 ≤ Days with lightning	60	Stop running	50

Table 8.7 Mud level parameters in geological early warning system

Level	Mud level/h		Speed limit/(km/h)	
	Mud level interval	Early warning threshold	Speed range	Early warning threshold
Level 1	0 < H ≤ Normal flood level	–	300 < Speed ≤ 350	350
Level 2	Normal flood level < H < 0.8 h	Normal flood level	250 < Speed ≤ 300	300
Level 3	0.8 h ≤ H < 0.9 h	0.8 h	200 < Speed ≤ 250	250
Level 4	0.9 h ≤ H < 1.1 h	0.9 h	150 < Speed ≤ 200	200
Level 5	1.1 h ≤ H < 1.3 h	1.1 h	100 < Speed ≤ 150	150
Level 6	1.3 h ≤ H < 1.5 h	1.3 h	50 < Speed ≤ 100	100
Level 7	H ≥ 1.5 h	1.5 h	Stop running	50

8.4.2 HSR Integrated Early Warning System for Natural Disasters

HSR integrated early warning system for natural disasters plays a positive role in reducing the impact of natural disasters on the HSR safe operation. Especially with the development of the technology, the HSR integrated early warning system for natural disasters will also have increasingly rich and comprehensive contents, so as to better respond to the threat of natural disasters to the HSR safe operation.

The HSR integrated early warning system for natural disasters mainly includes the wind warning system, rain warning system, snow warning system, earthquake warning system, lightning warning system and geological hazard warning system and other subsystems. The HSR integrated early warning system for natural disaster is also an important technical means to ensure the HSR safe operation. HSR integrated early warning system for natural disasters has main functions including real-time monitoring function, prompt function of alarms and speed limit, emergency response function and statistical query function, etc. to realize data exchange, interconnection and interoperability, and data sharing with the disaster monitoring and alarm systems of neighboring departments inside and outside the HSR, as shown in Fig. 8.11.

(1) **Architecture of the HSR natural disaster warning system** The disaster warning system based on a two-level architecture consists of the central system of the railway bureau of the Ministry of Transport and the on-site monitoring equipment. The on-site monitoring equipment consists of on-site data collection equipment and early warning systems. The central system of the railway bureau of the Ministry of Transport is composed of a data processing center and a front-end application. It collects and processes wind, rain, snow, lightning, geological disaster and earthquake monitoring and alarm information, and exchanges and shares information with the flood control and management system, the integrated

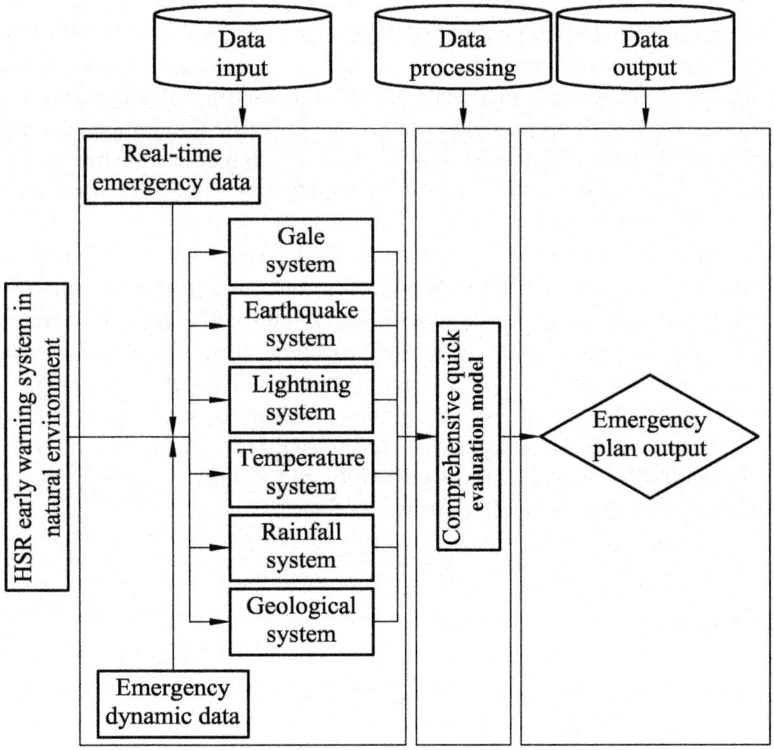

Fig. 8.11 Rapid assessment process of the HSR natural disaster early warning

video early warning system, other systems and the provincial meteorological bureau. In case of heavy rain, snow and foreign object intrusion alarm at the monitoring point, the central system of the railway bureau of the Ministry of Transport will send alarm to the integrated video early warning system.

(2) **Information flow of the HSR natural disaster warning.** The monitoring data and alarm data from the HSR integrated natural disaster warning system and off-road systems (including warning units, alarm processing equipment, adjacent railway bureaus) flow into the data processing center. After being parsed by the front-end processor, the monitoring data and alarm data flow to the database server, and the alarm data flow to the application server for alarm disposal. After the alarm disposal, the monitoring data and alarm data in the database finally flow to each user terminal, and the alarm information after the alarm disposal then flow to relevant systems.

8.5 Summary

The purpose of setting up the HSR early warning system for natural disasters is to improve the HSR safe operation. Therefore, the HSR early warning system for natural disasters must have high reliability. However, the reliability design of the HSR early warning system for natural disasters is not only reflected in the redundancy configuration of the system hardware, but also the setting of the alarm threshold value. The alarm threshold provided in this chapter is the recommended value, and the HSR alarm threshold which needs to be subject to dynamic adjustments at certain intervals according to the actual situation of the HSR operation, so as to achieve the purpose of the warning system.

In the HSR safe operation, the HSR integrated early warning system for natural disaster can realize data sharing, interconnection and other functions. The HSR integrated early warning system for natural disasters in the early stage of operation may have a high rate of false alarms, which requires us to have a correct understanding of the use of the HSR early warning system for natural disasters. We should not stop using the system when there are false alarms. With the system, we will be able to reduce damage caused by disasters. Therefore, the HSR integrated early warning system for natural disasters, with its reliability and stability improved, can protect the HSR safe operation more effectively.

Bibliography

1. Lin H, Haili L (2005) Analysis of flow characteristics around high-speed railway bridge-vehicle system under cross wind. J Southwest Jiaotong Univ 40(5):585–590
2. Zhifu M, Xiaobin Z, Yunjuan M et al (2010) Research on strong crosswind and crosswind protection measures and countermeasures on the Beijing-Shanghai high-speed railway (Jingxu section) [R]. Academy of Macroeconomic Research, Beijing

3. Yunjuan M, Shuhong M (2009) Research on the design wind speed of bridges on my country's high-speed railway passenger dedicated line. Railway Qual Control 18(10):34–37
4. Tian Z, He X, Weiwei G (2015) Running safety analysis of the train considering sudden change of wind load induced by wind barrier. J Vibr Eng 28(1):122–128
5. Haozhi P, Cong C (2020) How do high-speed rail projects affect the agglomeration in cities and regions. Transp Res Part D Transp Environ 88:102561
6. Li T (2022) Spatiotemporally complementary effect of high-speed rail network on robustness of aviation network. Transp Res Part A Policya Pract 155(1):95–114
7. Kuan Q, Yanliang D, Fugui Q (2008) Train crosswind and strong wind countermeasure research in Japan. J China Railway Soc 30(1):82–88
8. Jianshe S (2009) Stability analysis of train overturning in crosswind and design of automatic monitoring system for railway windbreak warning. J Zhengzhou Railway Vocat Tech Coll 21(1):5–12
9. Zhengyu Q (2009) Railway gale disaster in Northwest China and its prevention and control countermeasures. Chin Railways 47(3):1–4
10. Shuli L, Yulai W, Hun-Koo H (2019) et al (2019) Impact of high-speed rail network development on airport traffic and traffic distribution: Evidence from China and Japan. Transp Res Part A Policy Pract 127:115–135

SPRINGER NATURE

GPSR Compliance

The European Union's (EU) General Product Safety Regulation (GPSR) is a set of rules that requires consumer products to be safe and our obligations to ensure this.

If you have any concerns about our products, you can contact us on ProductSafety@springernature.com

In case Publisher is established outside the EU, the EU authorized representative is:

Springer Nature Customer Service Center GmbH
Europaplatz 3
69115 Heidelberg, Germany

The manufacturer's authorised representative in the EU is Springer Nature Customer Service Centre GmbH, Europaplatz 3, 69115 Heidelberg, Germany. If you have any concerns regarding our products, please contact ProductSafety@springernature.com

Printed and bound by CPI Group (UK) Ltd, Croydon, CR0 4YY
01/10/2026
02225867-0004